Statistics in Natural Resources

To manage our environment sustainably, professionals must understand the quality and quantity of our natural resources. Statistical analysis provides information that supports management decisions and is universally used across scientific disciplines. *Statistics in Natural Resources: Applications with R* focuses on the application of statistical analyses in the environmental, agricultural, and natural resources disciplines. This is a book well suited for current or aspiring natural resource professionals who are required to analyze data and perform statistical analyses in their daily work. More seasoned professionals that have previously had a course or two in statistics will also find the content familiar. This text can also serve as a bridge between professionals who understand statistics and want to learn how to perform analyses on natural resources data in R.

The primary goal of this book is to learn and apply common statistical methods used in natural resources by using the R programming language. If you dedicate considerable time to this book, you will:

- Develop analytical and visualization skills for investigating the behavior of agricultural and natural resources data.

- Become competent in importing, analyzing, and visualizing complex data sets in the R environment.

- Recode, combine, and restructure data sets for statistical analysis and visualization.

- Appreciate probability concepts as they apply to environmental problems.

- Understand common distributions used in statistical applications and inference.

- Summarize data effectively and efficiently for reporting purposes.

- Learn the tasks required to perform a variety of statistical hypothesis tests and interpret their results.

- Understand which modeling frameworks are appropriate for your data and how to interpret predictions.

- Includes over 130 exercises in R, with solutions available on the book's website.

Matthew Russell is a forest analytics consultant at Arbor Custom Analytics LLC where he uses data to solve natural resources problems. He is the author/co-author of more than 75 peer-reviewed publications focused on applied forestry research. He has conducted extensive research and teaching on topics related to forest modeling and statistics. He regularly offers short courses and workshops on data science and R for natural resources and environmental professionals.

Statistics in Natural Resources

Resources

Applications with R

Matthew Russell

CRC Press
Taylor & Francis Group
Boca Raton London New York

CRC Press is an imprint of the
Taylor & Francis Group, an **informa** business

A CHAPMAN & HALL BOOK

First edition published 2023
by CRC Press
6000 Broken Sound Parkway NW, Suite 300, Boca Raton, FL 33487-2742

and by CRC Press
4 Park Square, Milton Park, Abingdon, Oxon, OX14 4RN

CRC Press is an imprint of Taylor & Francis Group, LLC

Library of Congress Cataloging-in-Publication Data

Names: Russell, Matthew B. (Computer scientist) author.
Title: Statistics in natural resources : applications with R / Matthew Russell.
Description: 1 Edition. | Boca Raton, FL : CRC Press, 2023. | Includes
bibliographical references and index.
Identifiers: LCCN 2022009570 (print) | LCCN 2022009571 (ebook) | ISBN
9781032258782 (hardback) | ISBN 9781032259543 (paperback) | ISBN
9781003285809 (ebook)
Subjects: LCSH: Natural resources--Statistical methods. | R (Computer
program language) | Data mining.
Classification: LCC HC59.3 .R87 2023 (print) | LCC HC59.3 (ebook) | DDC
330.01/5195--dc23/eng/20220401
LC record available at https://lccn.loc.gov/2022009570
LC ebook record available at https://lccn.loc.gov/2022009571

ISBN: 978-1-032-25878-2 (hbk)
ISBN: 978-1-032-25954-3 (pbk)
ISBN: 978-1-003-28580-9 (ebk)

DOI: 10.1201/9781003285809

Typeset in Latin modern
by KnowledgeWorks Global Ltd.

Publisher's note: This book has been prepared from camera-ready copy provided by the authors.

To natural resources students. Past, present, and future.

Contents

Preface

0.1 What this book is about

The primary goal of this book is to learn and apply common statistical methods used in natural resources by using the R programming language. This book encompasses applied and theoretical techniques commonly used in these disciplines. A key component of the book is learning how to make inference with diverse data sources using R.

To manage our environment sustainably, professionals must understand the quality and quantity of our natural resources. Statistical analysis provides information that supports management decisions and is universally used across scientific disciplines. This book focuses on the application of statistical analyses in the environmental, agricultural, and natural resources disciplines.

Chapters 1 through 10 form the basis of most semester-long introductory statistics classes at the undergraduate level. Chapters 11 through 15 could also be included in an introductory statistics class for graduate students that moves at an accelerated pace.

If you dedicate considerable time to this book, you do the following:

- Develop analytical and visualization skills for investigating the behavior of agricultural and natural resources data.
- Become competent in importing, analyzing, and visualizing complex data sets in the R environment.
- Recode, combine, and restructure data sets for statistical analysis and visualization.
- Appreciate probability concepts as they apply to environmental problems.
- Understand common distributions used in statistical applications and inference.
- Summarize data effectively and efficiently for reporting purposes.
- Learn the tasks required to perform a variety of statistical hypothesis tests and interpret their results.
- Understand which modeling frameworks are appropriate for your data and how to interpret predictions.

0.2   What this book is not about

This book does not cover how natural resources data are collected and/or sampled. For the purposes of learning, we will mostly work with "tidy" data sets that have been collected by others. Many other texts in the disciplines of environmental sampling and experimental design are available that cover these topics in more depth. You would do well to have one of these courses as a part of your quantitative expertise.

The discipline of data science[1] has emerged in the last decade with a number of associated quantitative methods, such as machine learning, support vector machines, and unsupervised learning. Data science is an interdisciplinary field that often uses statistics in combination with computer science and a deep knowledge within a domain of interest. This book will focus instead on the inference one can make by performing statistical tests. As Blei and Smyth (2017) describe, the field of data science differs from statistics in many ways, but also has some similarities.

0.3   Prerequisites

You will get the most out of this book if you have already completed an undergraduate course in statistics or have comparable quantitative skills. If you can explain the difference between a mean and median, understand the basic concepts behind linear regression, and can describe what a two-sample *t*-test seeks to accomplish, you will be well prepared for the concepts in this book. There are a large number of textbooks, blogs, and online resources that cover these topics in depth. This book will also cover some of those topics, but will present them in more depth.

To a lesser degree, experience in programming is a plus but not required. A member of my PhD committee commented to me once that many universities counted programming courses as satisfying a foreign language requirement for students in the late 20th century. While diminishing to the world's spoken word, this observation reflects the value of programming skills in today's workforce. As it relates to R, books such as *Hands-On Programming with R*[2]

[1]https://en.wikipedia.org/wiki/Data_science
[2]https://rstudio-education.github.io/hopr/

by Garrett Grolemund and *R for Data Science*[3] by Hadley Wickham and Grolemund are excellent places to start.

Historically, many students learned about the concepts of statistics in a lecture format by doing calculations on paper with small data sets. Often in a separate lab component of a course, students learned how to use software to apply the theoretical concepts learned through lectures. So, to excel in a statistics class, students were required to learn both the theory and application of statistical concepts. Today, software such as R (and more specifically the **tidyverse** suite of packages), provide a more seamless integration of both the theory and application of statistical concepts.

To get started, you will need to do three things:

1. **Download R.** R is one of the most popular programs for learning statistics. In short, R is a programming language and software environment for statistical computing and graphics. Download R at the Comprehensive R Archive Network (CRAN)[4] and choose your operating system.

2. **Download RStudio.** RStudio is an integrated development environment that provides an interface to R. If it were a car, R would be the engine that moves it and RStudio would be the dashboard where the driver controls the wheel and can see how the car is performing. In short, RStudio allows us to efficiently and effectively work with data using the R system. Download RStudio Desktop at the RStudio products webpage[5]. If you are hesitant to jump right into the book or have issues with installing programs, consider using RStudio Cloud[6], a web-based interface to RStudio. A free version allows you to begin 15 projects with up to 15 usage hours per month.

3. **Install packages**. In the R program, packages are a collection of functions and data sets written by R users. You can perform a lot of analyses using R "off the shelf," or what is termed base R. There are tens of thousands of packages archived by CRAN. For perspective, 31 of these packages were developed by forestry professionals for specific applications within that discipline (Russell 2020).

---

[3]https://r4ds.had.co.nz/
[4]https://cran.r-project.org/
[5]https://rstudio.com/products/rstudio/
[6]https://rstudio.cloud/

0.3.1 Installing R packages

The **tidyverse** package is a "megapackage" that includes several packages that import, reshape, and visualize data in a consistent manner, among other tasks. Install the tidyverse package with the following line of code:

```
install.packages("tidyverse")
```

After installing the package, you will need to load it using the `library` command to use its functions:

```
library(tidyverse)
```

Interestingly, few statistical functions are available in the **tidyverse** package. So why spend the time learning about it? The tidyverse has emerged as an excellent foundation to learn about statistics through its philosophy of organizing and manipulating data (Wickham et al. 2019). We will use the tidyverse heavily to import, wrangle, and visualize data.

All of the data sets used in this book can be accessed by installing **stats4nr**, an R package developed for this book. First, you'll need to install the **devtools** package:

```
install.packages("devtools")
```

You can then install the package that contains the data sets for this book by installing it through GitHub:

```
devtools::install_github("mbrussell/stats4nr")
```

You can load the package using `library()` and load a data set by typing the name of it. For example, **ant** is the name of one of the data sets containing information on ant species richness in bogs and forests in New England, USA. We can use the `head()` function in R to print the first six rows in the data:

```
library(stats4nr)
head(ant)
```

```
##   site ecotype spprich   lat elev
## 1  TPB  Forest       6 41.97  389
## 2  HBC  Forest      16 42.00    8
## 3  CKB  Forest      18 42.03  152
## 4  SKP  Forest      17 42.05    1
## 5   CB  Forest       9 42.05  210
## 6   RP  Forest      15 42.17   78
```

A brief description and sources of all data sets in this book can be found on GitHub: https://github.com/mbrussell/stats4nr

0.4 Working with data sets

There are many formats to store data in R such as vectors, matrices, or lists. Nearly all of our data sets used in this book will be stored as data frames, or what the **tidyverse** terms "tibbles." The data frame is analogous to how you would view data organized in a spreadsheet: a column contains values of a variable and a row contains values from each column.

At appropriate times in the book we will install and use other data sets and packages to perform specific statistical tasks.

0.5 Base R versus the tidyverse

The collection of statistical functions available in base R are numerous and we will rely on these heavily. An excellent introduction to many of these base R functions can be found in Dalgaard (2008). We will also use the **tidyverse** suite of functions to help recode and structure data to bring it to a format that can be analyzed. The Wickham and Grolemund[7] text is an excellent source for learning these **tidyverse** functions. In our application of these methods in this text, it will not be uncommon to use a base R function within a **tidyverse** function to perform a statistical analysis.

---

[7]https://r4ds.had.co.nz/

0.6   Conventions with text and font

Understanding how certain text and fonts appear in the book can aid you in understanding the concepts:

- Words that are in **bold** indicate new concepts, names of R packages, or names of data sets.
- Words that appear in `constant width` indicate elements of a data set such as a variable name.
- Words that appear in `constant width` followed by a parentheses `()` indicate a function that performs an operation.

This version of the book was built with R version 4.1.0 (2021-05-18).

0.7   Acknowledgments

I am indebted to the countless colleagues, professors, and students that have been with me along my journey in learning statistics. This includes faculty, staff, and students from the University of Minnesota, University of Maine, and Virginia Tech. I am also thankful to the community of officers and members of the Society of American Foresters A1-Inventory and Biometrics Working Group.

I am particularly indebted to the graduate students of the University of Minnesota's Natural Resources Science and Management program for their desire for a graduate-level statistics class offered through the program. Were it not for my time spent with graduate students during my faculty interview in January of 2014, a lunch hour pizza session with Q&A, I may not have known the true importance of offering quantitative classes from within the discipline. Thank you students.

Finally, a special appreciation to my wife who lent her support and encouragement throughout the pandemic while I toiled on this project. Thanks, Annie.

0.8   References

Blei, D.M., P. Smyth. 2017. Science and data science. *Proceedings of the National Academies of Science* 114(33): 8689–8692.

Grolemund, G. 2014. *Hands-On programming with R: write your own functions and simulations.* O'Reilly Media. 230 p. Available at: https://rstudio-education.github.io/hopr/index.html

Wickham, H., and G. Grolemund. 2017. *R for data science: import, tidy, transform, visualize, and model data.* O'Reilly Media. 520 p. Available at: https://r4ds.had.co.nz/

Russell, M.B. 2020. Nine tips to improve your everyday forest data analysis. *Journal of Forestry* 118: 636–643.

Wickham, H., et al. 2019. Welcome to the tidyverse. *Journal of Open Source Software* 4(43): 1686. Available at: https://doi.org/10.21105/joss.01686

1

Visualizing data

1.1 Introduction

Good statistical analyses begin and end with visualizations. By visualizing your data before conducting statistical analyses, you will discover patterns and identify interesting observations in your data. Many of the techniques involved in exploring and acquainting yourself with data have been pioneered in the field of exploratory data analysis (Tukey 1977).

There are several parts to the data analysis life cycle, and visualizing data should be done frequently during this process:

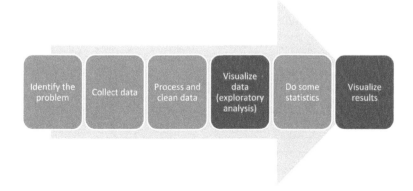

FIGURE 1.1 The data analysis life cycle.

In this chapter we will learn how to describe different components of data. This will make visualizing the data much easier because we will use the terminology that is appropriate for the variables we are working with. After we visualize the data, this usually gives us a good indication about which kinds of statistical analyses can be done with our data.

To implement the visualizations in the chapter, we will use the **ggplot2** package, one of the core packages found in the **tidyverse** megapackage. The *gg* in **ggplot2** is an abbreviation of "grammar of graphics," a language that provides insight into the deep structure behind visualizations. Resources developed by Wilkinson (2005) and Wickham (2010) provide much more on the theory and application of this language.

If this is your first time using the **tidyverse** package, you will need to install it:

```
install.packages("tidyverse")
```

Then, you will need to load the package using the `library()` function:

```
library(tidyverse)
```

We will use the **ggplot2** functions extensively throughout this chapter to learn about data through visualizations.

Visualizing and graphing data are grouped in the area of descriptive statistics. It is important to learn these skills so that we can inform our statistical tasks in later chapters. And because all good analyses end with visualization, our last chapter will focus on how we design graphs and visualizations to clearly convey our statistical results.

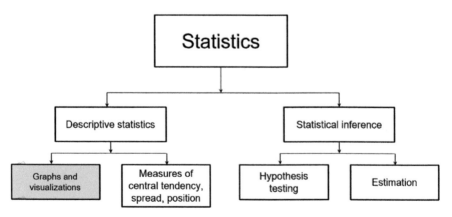

FIGURE 1.2 This chapter focuses on how graphs and visualizations are used in descriptive statistics.

1.2 Components of data

In the simplest sense, data are a collection of facts. In natural resources disciplines, data are often collected across long time periods and at multiple resolutions. Increasingly, natural resources professionals are integrating data they collect with other data. For example, a study about the productivity of forests might integrate both forest inventory data with current and future climate data to understand the effects of global change patterns on tree growth (e.g., Weiskittel et al. 2011).

To be effective in visualizing data, it will help if we understand how to describe the various components of data. First, data need to be organized in a structured format to get the most value from them. Structuring data in a *tidy* format facilitates this, that is, where every variable is a column, every observation is a row, and each type of observational unit is contained in a table (Wickham 2014). Keeping data tidy often involves creating "rectangles" of data within a spreadsheet-like format (Broman and Woo 2018). Natural resources data can also be unstructured, such as documents with large amounts of text from transcripts or environmental remote sensing data with voluminous pixels, but data will need to be wrangled to transition it into a usable format.

In R, most natural resources data are best analyzed by importing and working with data as **data frames**. Within these data frames, **variables** are stored as columns and **observations** (also defined as cases or records) as rows. When we visualize data, the variable names are plotted along the axes of graphs and the observations make up the elements of a graph.

1.2.1 Categorical variables

A **categorical variable** (also known as a factor) places an observation into a group or category. Examples include season of the year (i.e., spring, summer, fall, or winter) and plant and animal taxonomy (e.g., genus and species). Categorical variables use a **nominal** approach to label observations that fit into a group.

The number of categories that a variable can take can be numerous. Generally, as the number of categories for a variable increases, it will be more difficult to visualize and test for differences across the categories. At the other extreme, a categorical variable can be **binary** in which it takes only one of two possible outcomes. Examples include presence or absence, alive or dead, and positive or negative. While categorical variables are not necessarily quantifiable, if a categorical variable is **ordinal** it indicates that the order of values is important, but the difference between each order cannot be quantified.

1.2.2 Quantitative variables

Quantitative variables take numerical values. These variables can be **discrete**, where data are based on integers or counts. An example of discrete data includes the number of plant species found within a genus. Discrete variables are common in natural resources where they are often referred to as **count data**.

Quantitative variables can also be **continuous** where they take on any value within an interval. An example is the current air temperature. Generally, if you can add, subtract, multiply, and divide a variable by another variable, it has the properties of a quantitative variable and is not categorical.

DATA ANALYSIS TIP: Oftentimes categorical variables are coded as quantitative variables in a data set. For example, a study on the biology of deer may code its sex as a 1 (male) or 0 (female). This may not matter much in the data exploration stage (although you'll need to know which sex each number represents), but it will have tremendous impacts at the stage of statistical analysis. It is a good practice to recode these "categorical variables in disguise" so that R recognizes them as categorical variables, e.g., "Male" and "Female."

1.2.3 The elm data set

The **elm** data set in the **stats4nr** package contains several variables that are useful for understanding the components of data. These data contain observations on 333 cedar elm trees (*Ulmus crassifolia* Nutt.) measured in Austin, Texas (Russell 2020). We will load the data as a data frame and then print out to the R console:

```
library(stats4nr)
elm
```

FIGURE 1.3 Leaves and flowers of a cedar elm tree (photo: W.D. Brush, USDA-NRCS PLANTS Database).

```
## # A tibble: 333 x 8
##    STATUSCD  SPCD   DIA    HT CROWN_HEIGHT CROWN_DIAM_WIDE UNCOMP_CROWN_RATIO
##       <dbl> <dbl> <dbl> <dbl>        <dbl>           <dbl>              <dbl>
## 1        1   838     5    32         19.2              19                 60
## 2        1   838     5    25         11.3              11                 45
## 3        1   838   5.1    21          6.3              10                 30
## 4        1   838   5.1    27         18.9              13                 70
## 5        1   838   5.1    22         18.7               6                 85
## 6        1   838   5.1    27         18.9              11                 70
## 7        1   838   5.2    29         11.6              12                 40
## 8        1   838   5.2    20          7                 9                 35
## 9        1   838   5.2    18         17.8              12                 99
## 10       1   838   5.2    17          6                17                 35
## # ... with 323 more rows, and 1 more variable: CROWN_CLASS_CD <dbl>
```

By default, R will print the first 10 rows of a data frame in the *tibble* format. Tibbles are essentially data frames used within the **tidyverse** package. Note that R reads in the variables in the col_double() format, or a quantitative variable that is not an integer.

The variables in **elm** include:

- DIA, the tree's diameter at breast height, measured in inches,
- HT, the total height of the tree, measured in feet,
- CROWN_HEIGHT, the height at the base of the crown, measured in feet,

- CROWN_DIAM_WIDE, the width of the live crown at the widest point, measured in feet, and
- CROWN_CLASS_CD, the tree's crown class code that indicates the relative crown position of the tree: Open grown (1), Dominant (2), Co-dominant (3), Intermediate (4), or Suppressed (5).

A few handy functions allow you to inspect the contents of any data frame. The dim() function returns the number of observations and variables, or dimensions in a data frame:

```
dim(elm)
```

```
## [1] 333    8
```

head() and tail() return the first and last six lines of observations, respectively:

```
head(elm)
```

```
## # A tibble: 6 x 8
##   STATUSCD  SPCD   DIA    HT CROWN_HEIGHT CROWN_DIAM_WIDE UNCOMP_CROWN_RATIO
##      <dbl> <dbl> <dbl> <dbl>        <dbl>           <dbl>              <dbl>
## 1       1   838   5      32         19.2              19                 60
## 2       1   838   5      25         11.3              11                 45
## 3       1   838   5.1    21          6.3              10                 30
## 4       1   838   5.1    27         18.9              13                 70
## 5       1   838   5.1    22         18.7               6                 85
## 6       1   838   5.1    27         18.9              11                 70
## # ... with 1 more variable: CROWN_CLASS_CD <dbl>
```

```
tail(elm)
```

```
## # A tibble: 6 x 8
##   STATUSCD  SPCD   DIA    HT CROWN_HEIGHT CROWN_DIAM_WIDE UNCOMP_CROWN_RATIO
##      <dbl> <dbl> <dbl> <dbl>        <dbl>           <dbl>              <dbl>
## 1       1   838  23.7    37         27.8              53                 75
## 2       1   838  24.7    52         44.2              42                 85
## 3       1   838  28.1    39         37.1              48                 95
## 4       1   838  28.7    62         55.8              55                 90
## 5       1   838  29      46         27.6              45                 60
## 6       1   838  43      70         66.5              50                 95
## # ... with 1 more variable: CROWN_CLASS_CD <dbl>
```

The summary() function provides summary statistics for all quantitative variables in the data, including the mean, median, and quantile values:

```
summary(elm)
```

```
##     STATUSCD      SPCD          DIA             HT          CROWN_HEIGHT
## Min.   :1   Min.   :838   Min.   : 5.00   Min.   :15.00   Min.   : 4.8
## 1st Qu.:1   1st Qu.:838   1st Qu.: 6.60   1st Qu.:24.00   1st Qu.:15.0
## Median :1   Median :838   Median : 8.80   Median :30.00   Median :18.9
## Mean   :1   Mean   :838   Mean   :10.42   Mean   :31.53   Mean   :20.3
## 3rd Qu.:1   3rd Qu.:838   3rd Qu.:12.70   3rd Qu.:37.00   3rd Qu.:24.7
## Max.   :1   Max.   :838   Max.   :43.00   Max.   :70.00   Max.   :66.5
## CROWN_DIAM_WIDE UNCOMP_CROWN_RATIO CROWN_CLASS_CD
## Min.   : 4.00   Min.   :15.0       Min.   :1.000
## 1st Qu.:15.00   1st Qu.:50.0       1st Qu.:3.000
## Median :20.00   Median :65.0       Median :3.000
## Mean   :23.41   Mean   :64.8       Mean   :3.174
## 3rd Qu.:30.00   3rd Qu.:80.0       3rd Qu.:3.000
## Max.   :57.00   Max.   :99.0       Max.   :5.000
```

You might notice that the summary() function does not provide much detail on the categorical variable for CROWN_CLASS_CD. For categorical variables, the table() function works well and provides the number of observations for each category. You can follow this by using prop.table() to calculate each category's proportion of observations relative to the entire data set.

We can call any variable from a data frame by typing *dataframe$variable*. To calculate the number and proportion of observations in the **elm** data, we can use:

```
n_Crowns <- table(elm$CROWN_CLASS_CD)

n_Crowns
```

```
##
##   1   2   3   4   5
##   4   6 269  36  18
```

```
prop.table(n_Crowns)
```

```
##
```

```
##              1            2            3            4            5
## 0.01201201 0.01801802 0.80780781 0.10810811 0.05405405
```

As the data show, over 80% of the cedar elm trees in Austin, Texas have a co-dominant crown class.

1.2.4   Exercises

1.1 In your own discipline, find a data set or experiment that you're familiar with and reflect on the variables contained in the data. For categorical variables, list which variables are binary or ordinal. For quantitative variables, list which ones are discrete or continuous.

1.2 R has several built-in data sets that can be explored. Load **CO2**, a data set containing carbon dioxide uptake in plant grasses, by typing `CO2 <- tibble(CO2)`. Learn about the variables in the data by typing `?CO2`. Inspect the data and report the minimum and maximum values for the `uptake` variable. Determine how many plants were measured in the experiment and the number of chilled observations in the `Treatment` variable.

1.3 We can create new variables in existing data frames by using the `mutate()` function from the **dplyr** package. To add a new variable to an existing data frame, we can type the name of the data frame followed by "the pipe," written as `%>%` (or sometimes `|>`). The pipe is shorthand for saying "then." In other words, use my data frame "then" make a new variable in it.

As an example, we might be interested in making a new variable in the **elm** data that converts the diameter in inches to centimeters. To accomplish this, we would type `elm %>% mutate(DIA_cm = DIA * 2.54)`. The result is a new column called `DIA_cm` that contains the tree's diameter at breast height in centimeters. Multiple pipes can be written in a block of code, which we'll see later in this book.

Add a new variable to the **CO2** data frame, by collapsing the `conc` variable into a binary one. Research the `ifelse()` function and use it within the `mutate()` function to label all observations with an ambient carbon dioxide concentrations of 500 mL/L or greater as "HIGH" and all others as "LOW."

1.3 Graphics for visualizing data

Run the following code to plot data from the elm data set using the **ggplot2**
package. In this example, we will create a **scatter plot** showing the diameter
of elm trees on the x-axis and their height on the y-axis:

```
ggplot(data = elm, aes(x = DIA, y = HT)) +
        geom_point()
```

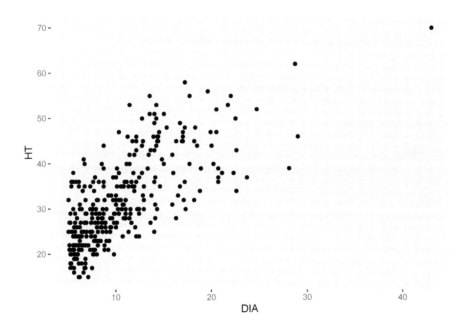

What results is a trend that we would expect—trees that are larger in diameter
are also taller. This is helpful because a tree's diameter is relatively easy to
measure, but a tree's height often requires more time and effort to obtain.

Now, we'll step through the code that produced the scatter plot above:

- The ggplot() function tells R that we want to produce a plot, and we need
  to tell it which data set and variables to use.
- The data = statement specifies that we want to plot variables from the **elm**
  data set.
- We specify the variables DIA and HT within the aes() statement. The aes is
  abbreviated for *aesthetics*, which allows us to change the properties of how

the data are shown in the graph. As it turns out, we have not done anything special with the current aesthetics in the scatter plot, but in the future we can add to the aesthetics by specifying different colors, shapes, and sizes to the data points.

- The geom_point() statement tells R that we want to produce a geometric object with points, i.e., a scatter plot. We will learn there are many "geom" types that can plot different layers of objects depending on the nature of the data and what you want to see.

Note that we add a + at the end of the line when we use ggplot(). This indicates we have more instructions to R before creating our scatter plot.

The way that we created the elm scatter plot will be the same style we'll use for all of our graphs. That is, the first line will tell ggplot() the data set and variables to plot (along with any additional aesthetics). The second line will specify which kind of graph to create (i.e., which "geom").

1.3.1   Visualizing categorical data

1.3.1.1   Bar plots

Bar plots are one of the most effective ways to display categorical data. These plots represent the categories as bars and their length shows the counts or percentages within each category. Before creating a bar plot, it is useful to investigate the values of the data in the plot. As an example, we may wish to analyze the number of trees in the **elm** data set by crown class codes. We can find the number of tree observations within each CROWN_CLASS_CD by using the count() function with a pipe:

```
elm %>%
    count(CROWN_CLASS_CD)
```

```
## # A tibble: 5 x 2
##    CROWN_CLASS_CD       n
##             <dbl> <int>
## 1               1     4
## 2               2     6
## 3               3   269
## 4               4    36
## 5               5    18
```

We observe that the greatest and fewest number of trees have a co-dominant and open grown crown class, respectively. We can also obtain the counts of observations with different code. The `count()` function makes this easy, but we also might want to determine the percentage of trees within each crown class category by grouping the data and then summarizing it.

To do this, we will start by grouping the data by `CROWN_CLASS_CD` using the `group_by()` statement. Then, we use the `summarize()` function to make a new variable called `n_trees` that sums the number of observations by `CROWN_CLASS_CD`. This step produces the same output that the `count()` function provides in the previous step, so it's not particularly novel. In the last line, we use the `mutate()` function to calculate a new variable `Pct` that provides the percentage of observations in the data within each crown class.

We'll also want to use this new data set in the future, so we will assign it a name `elm_summ`. These operations are handy tools found in the **dplyr** package, a core package in the **tidyverse** that allows you to transform and reshape data:

```
elm_summ <- elm %>%
    group_by(CROWN_CLASS_CD) %>%
    summarize(n_trees = n()) %>%
    mutate(Pct = n_trees / sum(n_trees) * 100)
elm_summ
```

```
## # A tibble: 5 x 3
##   CROWN_CLASS_CD n_trees   Pct
##            <dbl>   <int> <dbl>
## 1              1       4  1.20
## 2              2       6  1.80
## 3              3     269 80.8
## 4              4      36 10.8
## 5              5      18  5.41
```

We observe that the co-dominant and open grown crown classes represent 80.8% and 1.2% of the data, respectively.

1.3.1.2 Pie charts

Pie charts show the distribution of variables as a "pie" whose slices are sized by the counts or percentages for the categories. We can plot the elm tree crown class data as a pie chart, recalling the **elm_summ** data we created in an earlier step. Because we are plotting specific values in that data set, we'll specify `stat = "identity"` in the `geom_bar()` layer. The key to creating a pie chart is to specify `coord_polar()` in the last line. A pie chart shows the same information as a bar plot, but in polar coordinates:

```
ggplot(data = elm_summ, aes(x = "", y = n_trees,
                            fill = CROWN_CLASS_CD)) +
       geom_bar(stat = "identity") +
       coord_polar("y")
```

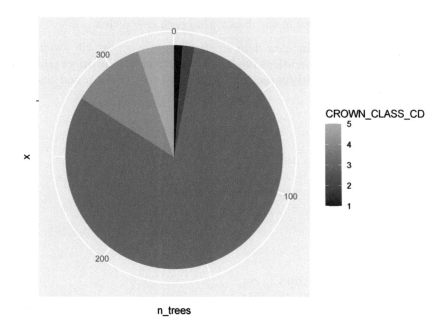

n_trees

Pie charts should be used with caution. There are several reasons for overlooking pie graphs and instead favoring other kinds of graphs. Several reasons behind this are pointed out in early work by Cleveland and McGill (1984, 1985):

- The human eye does a poor job in judging angles. This makes it difficult to discern the values within pie graphs.
- Pie graphs contain color, and the human eye does a very poor job in judging color in graphs. This is especially true for individuals that are color blind or have other visual impairments.

Regardless, because of the widespread use of pie graphs it is helpful to learn how to construct and interpret them. As an alternative, place more priority in developing graphs such as bar plots when displaying categorical data because the human eye does a good job of discerning position and lengths of objects.

1.3.1.3 Polar area graphs

Another use of polar coordinates in graphing is a **polar area graph**, also called a coxcomb plot or rose diagram. Florence Nightingale, a nurse and pioneer in statistical graphics, popularized the use of the polar area graphs in the mid-19th century. They are similar to pie charts, but they have identical angles and extend from the plot's center depending on the magnitude of the values that are plotted.

Think of the polar area diagram as a pie chart meets a histogram. A few advantages of polar area diagrams include:

- They are useful for plotting cyclical data. For example, the counts of a phenomenon in each of the 12 calendar months of a year.

- They are easy to read around the "rose" because data are presented chronologically.

- Multiple layers can be added within a diagram. Nightingale presented these kinds of layers in her visualizations of soldier deaths and wounds during the Crimean War.

In R, a polar area graph can be created by combining the geom_bar() and coord_polar() layers:

```
ggplot(data = elm, aes(x = CROWN_CLASS_CD)) +
        geom_bar() +
        coord_polar()
```

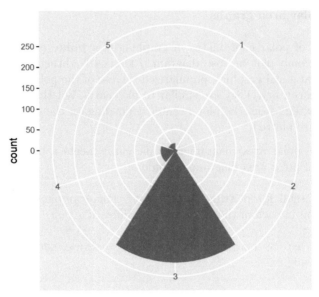

CROWN_CLASS_CD

For the elm data, the polar area graph reveals the large number of co-dominant trees. As you can see, polar area graphs have an advantage over pie charts because they allow the reader to see the depth within each category.

1.3.2   Visualizing quantitative data

1.3.2.1   Stem plots

Stem plots, also termed stem-and-leaf plots, are useful to separate each observation into a stem (all but the rightmost digit) and a leaf (the remaining digit). The process involves first positioning the stems in a vertical column, then drawing a vertical line to the right of the stems. The last step positions each leaf in the row to the right of its stem.

In R, the stem() function produces a stem plot and sorts its leaves ascending. Here is an example of a stem plot of the HT variable from the **elm** data set:

```
stem(elm$HT, width = 50)
```

```
##
##    The decimal point is 1 digit(s) to the right of the |
##
##    1 | 555667777777788888888999999
```

```
##    2 |  00000000001111111111111122222222222223+11
##    2 |  55555555555555555556666666666667777777+27
##    3 |  0000000000000000001111111222222222222
##    3 |  55555555555555555566666666666667777777+1
##    4 |  00000001122222333344444
##    4 |  5555555556666666677777888899
##    5 |  00011122333
##    5 |  55568
##    6 |  2
##    6 |
##    7 |  0
```

We can quickly observe that the greatest number of observations are between 25 and 29 feet tall.

1.3.2.2   Histograms and density plots

Histograms divide the possible values of a variable into classes or intervals of equal widths. The histogram shows how many observations fall within each interval. The height of each bar is equal to the number (or percent) of observations in its interval. The following code creates a histogram for HT:

```
ggplot(data = elm, aes(HT)) +
        geom_histogram()
```

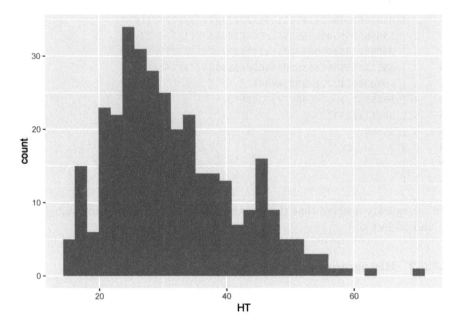

Any histogram can be reshaped depending on how many bins you specify in the `geom_histogram()` layer. For most applications, a bin width between 20 and 30 works well, but the appropriate number of bins will depend on the number of observations and the distribution of the data. Here is an example with 10 bins where you can see the coarser resolution that the histogram provides:

```
ggplot(data = elm, aes(HT)) +
      geom_histogram(bins = 10)
```

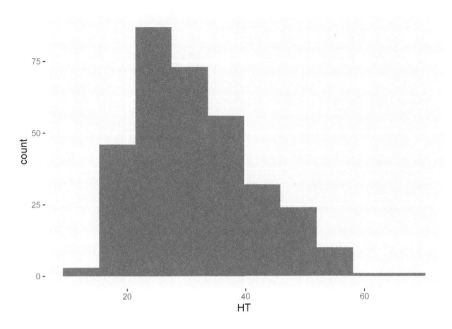

Density plots are similar to histograms but show the distribution of a variable using a smoothed curve. They may be advantageous over histograms because they show a detailed distribution and are not affected by the number of bins you select. The `geom_density()` layer provides density plots:

```
ggplot(data = elm, aes(HT)) +
        geom_density(color = "blue")
```

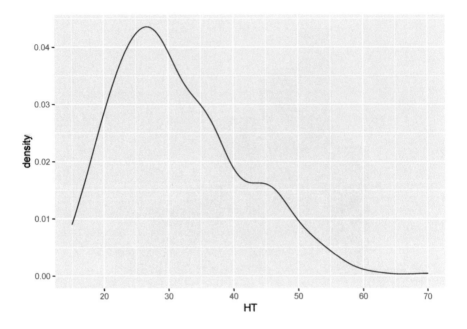

DATA ANALYSIS TIP We can describe the density or distribution of a variable as symmetric, left-skewed, or right-skewed. A distribution is symmetric if the right and left sides of the graph are approximately mirror images of each other. A distribution is right-skewed if the right side of the graph is much longer than the left side. A distribution is left-skewed if the left side of the graph is much longer than the right side. For the elm heights we can say that the distribution is right-skewed, indicating that there a few tall trees relative to many shorter trees.

1.3.2.3   Box plots and violin plots

The **median** of a variable and its **quartiles** divide the distribution roughly into quarters. The median is the middle value when a quantitative variable is sorted by its value. Three quartiles separate a variable into four parts, where the second quartile represents the median that separates the upper and lower half of observations. The first and third quartiles contain 25 and 75% of values below them, respectively. A **box plot** shows the minimum, first quartile (Q1), median, third quartile (Q3), and maximum values of a variable. Here is a box plot for the height of elm trees:

```
ggplot(data = elm, aes(x = 1, y = HT)) +
        geom_boxplot()
```

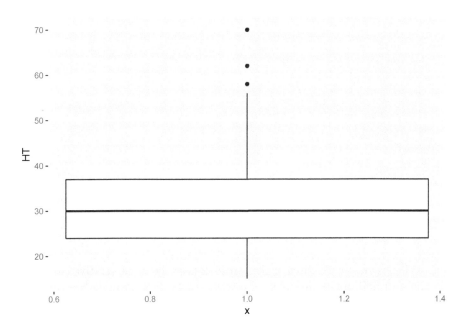

The Q1 and Q3 values make up the ends of the "box" in the box plot, while the minimum and maximum values make the "whiskers," leading to the alternative "box-and-whisker" plot name. You will also note that three observations are categorized as outliers. These outliers are greater than 55 feet tall and are shown as points. We will discuss how to deal with outliers in our statistical analyses later, but for now, you can understand that the tree observations are considered outliers because their magnitude sets them apart from the range of all the other observations.

Violin plots are similar to a box plot, but also show a kernel probability density of the data. Violin plots are quite similar to density plots. In a violin plot that shows the heights of the elm trees, we can see that the height peaks between 25 and 30 feet:

```
ggplot(data = elm, aes(x = 1, y = HT)) +
        geom_violin()
```

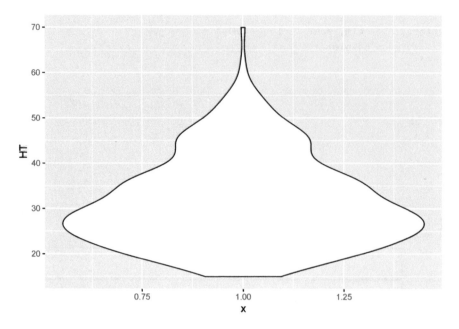

One of the greatest attributes of the **ggplot2** package is the relative ease that
you can add additional variables to existing plots. For example, we can add
the variable `CROWN_CLASS_CD` in the box plots and violin plots along the x-axis
to investigate the trends in height across different tree crown classes. Note that
although the crown class codes are numeric, we use `factor(CROWN_CLASS_CD)`
to tell R to treat them as categorical variables:

```
ggplot(data = elm, aes(x = factor(CROWN_CLASS_CD), y = HT)) +
        geom_boxplot()
```

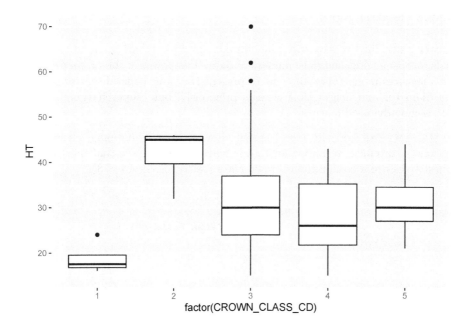

```
ggplot(data = elm, aes(x = factor(CROWN_CLASS_CD), y = HT)) +
        geom_violin()
```

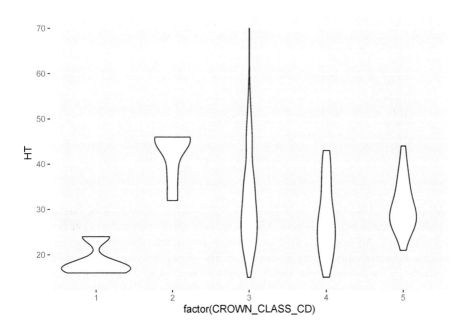

1.3.2.4 Scatter plots

Scatter plots, like we saw with the elm data previously, are some of the best
tools for visualizing bivariate numerical data. The power of the *aesthetics* and
geom features in `ggplot()` allow us to present the same general scatter plot of
elm diameter and height that we saw previously, but "supercharging" them
with some advanced features.

To start, we can add the tree's crown class code as a mapping variable in
the `aes()` statement to add color to our scatter plot. We see that most of the
tallest trees are a co-dominant crown class (`CROWN_CLASS_CD` = 3):

```
ggplot(data = elm, aes(x = DIA, y = HT,
                  color = factor(CROWN_CLASS_CD))) +
      geom_point()
```

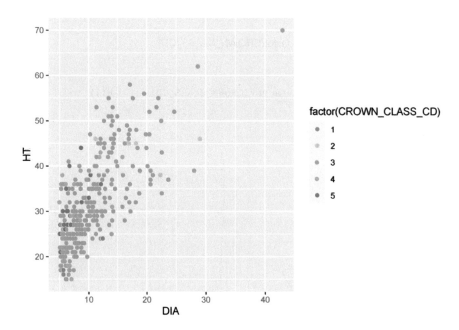

Adding a trend line can easily reveal a relationship between two continuous
variables. This is helpful if you need to make a quick approximation between
two variables. Adding a trend line can also reveal whether or not a linear or
nonlinear relationship exists in the data. The `geom_smooth()` can be added to
the code and will reveal the trends between the two variables. This function
fits a smoothed conditional mean to the data (in blue), along with confidence
intervals surrounding the estimate (in gray). After fitting a trend line to the

elm data you could say "A 20-inch diameter elm tree will be approximately 45 feet tall."

```
ggplot(data = elm, aes(x = DIA, y = HT)) +
        geom_point() +
        geom_smooth()
```

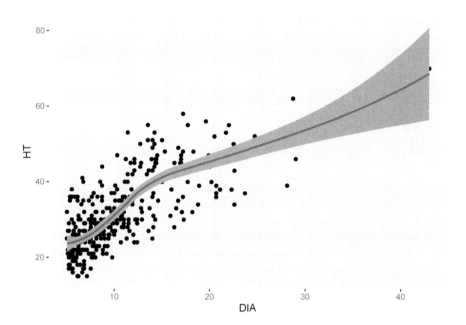

Another way to easily see the differences in ranges of two numerical variables in a scatter plot is to plot each level of a categorical variable in its own panel. The facet_wrap() statement allows you to do this.

In this case we easily see that co-dominant trees have a full range of DBH-HT, while the other crown classes have a narrower range with fewer observations:

```
ggplot(data = elm, aes(x = DIA, y = HT)) +
        geom_point() +
        facet_wrap(~CROWN_CLASS_CD)
```

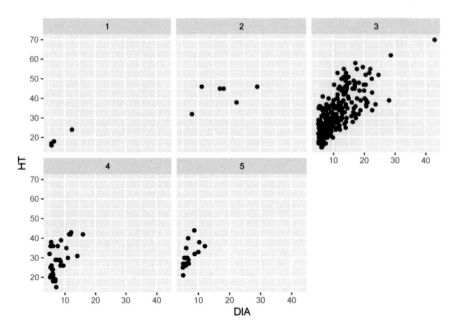

The `facet_wrap()` function works well when you have a single categorical variable to facet. The `facet_grid()` function allows you to plot two categorical variables simultaneously. We don't have another categorical variable in the elm data that could serve as a second variable. But you could imagine that if we had multiple tree species in the data set, we could plot the five crown classes vertically and the species horizontally.

A "hexagonal scatter plot" can be produced in `ggplot()`, which divides the x- and y-axes into hexagons, and the color of that hexagon reflects the number of observations in each hexagon. The `geom_hex()` layer fills in the number of observations within each hexagon. You can think of this as a scatter plot meets a histogram, where the colors indicate how many observations are contained within each bin.

The **hexbin** package in R provides functions to plot hexagonal scatter plots. Install the package first, and then load it to use the functions. Here's an example with the number of bins along the x- and y-axis set to 20:

```
install.packages("hexbin")
library(hexbin)
```

```
ggplot(data = elm, aes(x = DIA, y = HT)) +
        geom_hex(bins = 20)
```

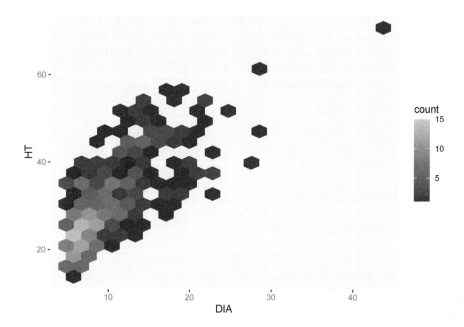

The hexagonal scatter plot shows that most of the observations in the elm data set are less than 10 inches in diameter and are shorter than 30 feet tall. In the original scatter plot, due to overlapping points in "busy" areas of the graph, this finding can't really be observed. Knowing that this pattern exists can be insightful for future data analysis.

The number of bins in `geom_hex()` can be increased to see a finer resolution (with fewer observations grouped into each hexagon). Or it can can be decreased to see a coarser resolution (with more observations grouped into each hexagon).

1.3.3   Exercises

1.4 Run the code `ggplot(data = elm, aes(x = DIA, y = HT))`. What is the result and why do you see what you see?

1.5 Using the **CO2** data set, write code using `ggplot()` that creates a bar plot displaying the number of measurements for each plant. Note that because the number of measurements are the same for each plant, the plot will look uniform.

1.6 Create a series of hexagonal scatter plots using the `uptake` variable in the **CO2** data set. Change the observations contained within each hexagonal bin to 5, 20, 40, and 60. (Remember to load the **hexbin** package.) What do you notice as the number of bins increases?

1.7 Create a grid of scatter plots by specifying the `facet_grid` statement in ggplot(). This statement creates a matrix of panels using two faceting variables. Plot the `conc` and `uptake` variables from the **CO2** data set within the scatter plot. Then, specify `Type` and `Treatment` as the faceting variables. Which `Type`, i.e., the location of the plants in the data set, contains the greater values of uptake?

1.4 Enhancing graphs

1.4.1 Adding text elements

Up to now, we've been using some of the basic plotting features available in **ggplot2**. Oftentimes these techniques are all that we need to complete our stage of exploratory data analysis. However, you will likely need to produce publication-quality graphs and figures to share with a broader audience as you continue your journey in data analysis. This section describes some additional code you can use in the ggplot() function to improve the quality of your graphs.

To start, we have already broken the cardinal sin in designing figures: we have not added units to the x- and y-axes to our plots. The labs() statement can be added, as we see here in the diameter and height trends from the elm data:

```
ggplot(data = elm, aes(x = DIA, y = HT)) +
       geom_point() +
       labs( x = "Tree diameter (inches)",
             y = "Tree height (feet)")
```

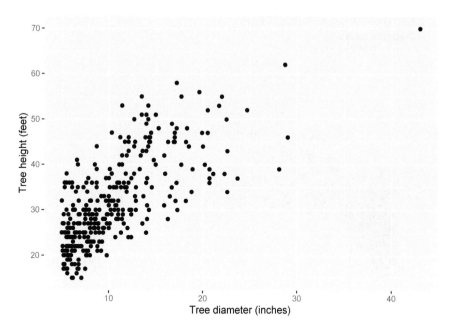

With `labs()` you can also add a title, subtitle, and caption to a graph. This can be helpful for describing the contents of the graph, a key result that the graph displays, and the source of the data:

```
ggplot(data = elm, aes(x = DIA, y = HT)) +
      geom_point() +
      labs( x = "Tree diameter (inches)",
            y = "Tree height (feet)",
            title = "Cedar elm trees in Austin, Texas",
            subtitle = "Trees that have a larger
            diameter are taller.",
            caption = "Source: USDA Forest Inventory and Analysis")
```

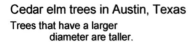

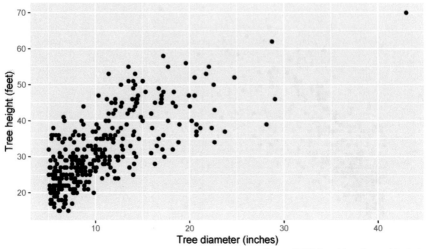

Source: USDA Forest Inventory and Analysis

You can also arrange the elements of a graph so that it is easier for the reader to understand. One of the key messages from the work of Cleveland and McGill (1984, 1985) is that trends are easily distinguished when they arranged as objects representing length along a common scale. This is helpful when displaying categorical variables such as bar plots. You can order elements of a graph ascendingly using the fct_infreq() statement. This function is available in the **forcats** package, another core set of functions in the **tidyverse** that deals with factor level variables (also known as categorical variables). Here is a bar plot of the number of trees by different crown classes from the elm data set with the values in ascending order:

```
ggplot(data = elm,
       aes(x = fct_infreq(factor(CROWN_CLASS_CD)))) +
       geom_bar()
```

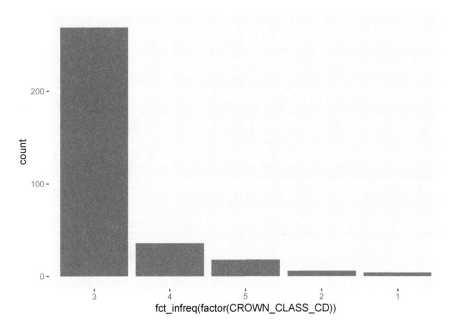

You can also use the `fct_rev()` function within **forcats** combined with `fct_infreq()` to sort the values descendingly:

```
ggplot(data = elm,aes(x = fct_rev(fct_infreq(factor(CROWN_CLASS_CD)))))+
      geom_bar()
```

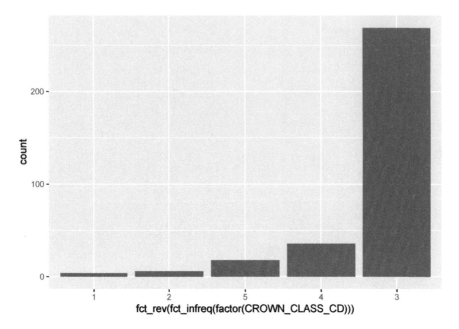

Natural resources data contain many ordinal variables. If a variable is ordinal in its design, it is effective in ordering the graph manually depending on the values of the variable. For example, because the tree crown classes depict how much sunlight a tree receives, we could reorder the elm data from the categories from open grown, dominant, dominant, intermediate, suppressed.

After exploring your data to understand trends, spending time to arrange the values and graphs so that they make more sense from a biological or numerical perspective will help the reader to better understand your analysis.

1.4.2   Adjusting plot layouts and themes

We have also been using the default ggplot output when looking at our graphs. A number of additional components can be adjusted to change the layout of a plot. A few examples include:

- The scale of x and y-axes can be adjusted with the statements `scale_x_continuous()` and `scale_y_continuous()`. The `limits =` statement can be specified here and allows you to change the upper and lower bounds of the axis.
- Legends can be repositioned to appear on the top, bottom, left, or right of the graph within the `theme()` statement.
- You can change the default style to a white background using `panel.background = element_rect(fill = "NA")`

You can continue to modify elements within the theme statement in `ggplot()`, however, there are also several default themes you can select. For example, here are the scatter plots for the elm data with the `theme_bw()`, `theme_classic()`, and `theme_void()` themes available in **ggplot2**:

```
ggplot(data = elm, aes(x = DIA, y = HT)) +
        geom_point() +
        labs(title = "theme_bw()") +
        theme_bw()
```

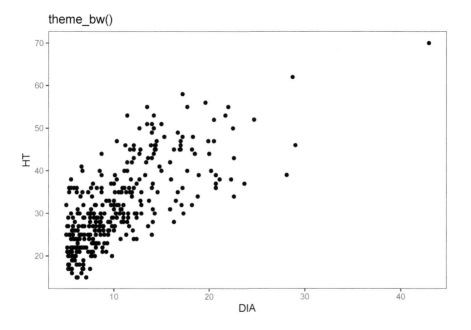

```
ggplot(data = elm, aes(x = DIA, y = HT)) +
        geom_point() +
        labs(title = "theme_classic()") +
        theme_classic()
```

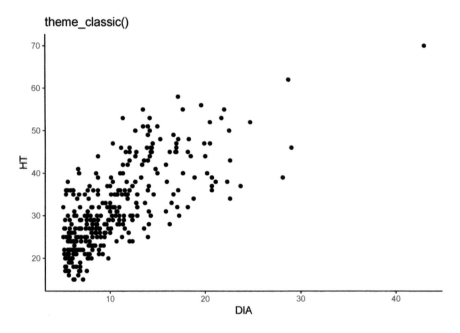

```
ggplot(data = elm, aes(x = DIA, y = HT)) +
        geom_point() +
        labs(title = "theme_void()") +
        theme_void()
```

theme_void()

You also may want to save your plot to your desktop to use in a report or to share on the web. The ggsave() function allows you to output a plot to disk. The ggsave() function allows you to create common image formats such as jpeg, tiff, png, and pdf. You can also change the width and height of the figure as an argument within ggsave(). Here's an example that will save a JPEG image named elm_scatter.jpg that saves a scatter plot of the elm data that is 3 inches in height and 5 inches width:

```
ggplot(data = elm, aes(x = DIA, y = HT)) +
        geom_point()

ggsave("elm_scatter.jpg", height = 3, width = 5, units = "in")
```

1.4.3 Exercises

1.8 Make a series of box plots for the uptake variable from the **CO2** data set. Each box plot should display the origin of the location of the plants in the data set, i.e., Quebec or Mississippi. Label the axes with the appropriate units and use the theme_bw() theme for your graph.

1.9 As you might have guessed, there is a "geom" for a bar plot and it's called geom_bar(). We discussed how the CROWN_CLASS_CD variable is an ordinal variable representing how much sunlight a tree receives. Create a bar plot showing the crown class codes on the x-axis and the number of observations on the y-axis using `ggplot()`.

1.10 Use `ggsave()` to save the graph you made in exercise 1.9 as a png image to your computer. Set the height and width of the figure to 10 cm and 10 cm, respectively.

1.11 Use `scale_x_continuous()` and `scale_y_continuous()` to "zoom in" on the scatter plot of elm diameter and height. Use the `limits` = statement to set the lower and upper bounds from 10 to 15 inches along the x-axis for diameter and 30 to 40 feet along the y-axis for height. Are there any trends in the data when looking at this region of the scatter plot?

1.5   Summary

Visualizing your data should be one of the first steps you take before doing any statistical analysis. Visualizing data allows you to better understand the data, spot any unusual trends or observations within your data, and provides you inspiration for deciding which kinds of statistical analyses might be appropriate for your data. It is important to be able to characterize the variables in your data. Whether data are categorical or quantitative, this will impact which types of visualizations are appropriate. For categorical data, bar charts, pie graphs, and polar area diagrams are tools to help visualize patterns. For quantitative data, histograms, box plots, and scatter plots are examples that work well. The **ggplot2** package is a core part of the **tidyverse** and provides a framework for visualizing data. There are a lot of options to supercharge your code for effective data visualization, but the components from this chapter will allow you to quickly visualize data so that we're better prepared to run statistical analysis in the following chapters.

1.6   References

Broman, K.W., and K.H. Woo. 2018. Data organization in spreadsheets. *American Statistician* 72(1): 2–10.

Cleveland, W.S., and R. McGill. 1984. Graphical perception: theory, experimentation, and application to the development of graphical methods. *Journal of the American Statistical Association* 79: 531–554.

Cleveland, W.S., and R. McGill. 1985. Graphical perception and graphical methods for analyzing scientific data. *Science* 229: 828–833.

Russell, M.B. 2020. Nine tips to improve your everyday forest data analysis. *Journal of Forestry* 118(6): 636–643.

Tukey, J.W. 1977. *Exploratory data analysis*. Addison-Wesley, Reading, MA. 712 p.

Weiskittel, A.R., N.L. Crookston, and P.J. Radtke. 2011. Linking climate, gross primary productivity, and site index across forests of the western United States. *Canadian Journal of Forest Research* 41: 1710–1721.

Wickham, H. 2010. A layered grammar of graphics. *Journal of Computational and Graphical Statistics* 19: 3–28.

Wickham, H. 2014. Tidy data. *Journal of Statistical Software* 59: 1–24. Available at: http://dx.doi.org/10.18637/jss.v059.i10

Wilkinson, L. 2005. *The grammar of graphics* (2nd ed.). *Statistics and Computing*. New York: Springer. 688 p.

2

Summary statistics and distributions

2.1 Introduction

Understanding a data set involves summarizing it so that you can better understand its characteristics. In combination with graphing and visualizing data, this chapter will complete our understanding of descriptive statistics.

This chapter will discuss descriptive statistics such as measures of **central tendency** (e.g., the mean, median, and mode), **spread**, (e.g., the variance and standard deviation), and **position** (e.g., quartiles and outliers). These values are often reported in final reports to convey the properties of data. We will round out the chapter by discussing distributions and **random variables**, numerical variables that describe the outcomes of a chance process. This will lead us into discussing probability in subsequent chapters.

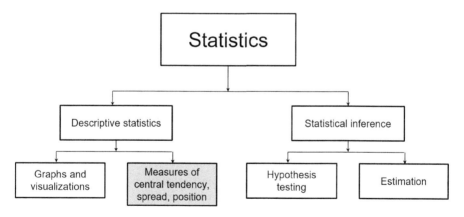

FIGURE 2.1 This chapter focuses on how summary statistics are used in descriptive statistics.

2.2   Summary statistics

A **statistic** can be defined as a summary of a sample. In contrast, a **parameter** represents a population constant. This section will discuss the common statistics we can calculate from samples. When we begin discussing random variables in the next section, you will begin to see how statistics are related to parameters.

2.2.1   Measures of central tendency

When a colleague mentions she calculated "the average" of a value, she is referring to a measure of central tendency. Measures of central tendency include the mean, median and mode.

The **mean**, denoted by $\bar{x}$, is calculated by summing all values and dividing by the number of values n:

$$\bar{x} = \frac{\sum_{i=1}^{n} x_i}{n} = \frac{x_1 + x_2 + \dots + x_n}{n}$$

The **median** is determined by finding the middle value of a sorted data set. The median is located in the $(n+1)/2$ position. The median is considered a *resistant* central tendency measure because compared to the mean, the median is not impacted by extreme values.

The **mode** is defined as the measurement that occurs most frequently. There are pros and cons to using these measures of central tendency, as indicated in Table 2.1.

As an example, we will determine various measures of central tendency on the number of chirps that a striped ground cricket makes. Ecologists believe that the number of chirps a striped ground cricket makes every second is related to the air temperature.

The data are from Pierce (1943) and contain the number of chirps per second (cps) and the air temperature, measured in degrees Fahrenheit (temp_F). The **chirps** data set from the **stats4nr** package contains the 15 observations of chirps per second:

```
library(stats4nr)
```

```
head(chirps)
```

TABLE 2.1 Pros and cons for various measures of central tendency.

Statistic	Pros	Cons
Mean	Most widely used in natural sciences	Affected by extreme values
Median	Not affected by extreme values; useful for skewed distributions	Does not take into account precise values
Mode	Effective if used with qualitative data	May not exist; can be more than one mode; may not be representative of the average for small samples.

```
## # A tibble: 6 x 2
##     cps temp_F
##   <dbl>  <dbl>
## 1   20    93.3
## 2   16    71.6
## 3 19.8    93.3
## 4 18.4    84.3
## 5 17.1    80.6
## 6 15.5    75.2
```

We can calculate the mean by hand for the number of chirps per second:

$$\bar{x} = \frac{20.0 + 16.0 + \dots + 14.4}{15} = 16.65$$

The median can be found by sorting the cps variable with the `arrange()` function. The median is located in the $(15+1)/2 = $ 8th position:

```
chirps %>%
  arrange(cps)
```

```
## # A tibble: 15 x 2
##      cps temp_F
##    <dbl>  <dbl>
## 1   14.4   76.3
## 2   14.7   69.7
## 3   15     79.6
## 4   15.4   69.4
```

```
## 5   15.5   75.2
## 6   16     71.6
## 7   16     80.6
## 8   16.2   83.3
## 9   17     73.5
## 10  17.1   80.6
## 11  17.1   82
## 12  17.2   82.6
## 13  18.4   84.3
## 14  19.8   93.3
## 15  20     93.3
```

The median chirps per second is 16.2. If the number of observations were even, we would calculate the average of the two middle observations in the sorted list. Fortunately, R has two built-in functions to calculate the mean and median:

```
mean(chirps$cps)
```

```
## [1] 16.65333
```

```
median(chirps$cps)
```

```
## [1] 16.2
```

Unfortunately, R does not have a built-in function to calculate a mode. Fortunately, the **modeest** package (Poncet 2019) contains the mfv() function, which returns the most frequent value (or values) in a numerical variable:

```
install.packages("modeest")
```

```
library(modeest)
```

```
mfv(chirps$cps)
```

```
## [1] 16.0 17.1
```

We see that there are two modes for the number of chirps per second: 16.0 and 17.1. These values are similar in magnitude to the mean and median.

2.2.2 Measures of spread

Aside from understanding the average of the data, or its *signal*, it is also informative to characterize how spread out the data are, or its *noise*. For example, the **range** , defined as the minimum subtracted from the maximum value, is a useful representation of how noisy the data are. The range is a quick measure of spread to calculate but is also very sensitive to extreme values.

The **variance** measures the average squared distance of the observations from their mean. To calculate the sample variance s^2, the average of the squared distance is determined:

$$s^2 = \frac{1}{n-1}\sum_{i=1}^{n}(x_i - \bar{x})^2 = \frac{1}{n-1}(x_1 - \bar{x})^2 + (x_2 - \bar{x})^2 + ... + (x_n - \bar{x})^2$$

While the variance is used widely in statistics, it is not always a meaningful number to characterize a variable of interest. This is because its units are squared. For example, we'll calculate the variance of the cps variable in the **chirps** data set :

$$s^2 = \frac{1}{15-1}(20.0 - 16.65)^2 + (16.0 - 16.65)^2 + ... + (14.4 - 16.65)^2 = 2.90$$

The calculation indicates that the variance is 2.90 squared chirps per second. For a more useful number, instead we'll take the square root of the variance and report the **standard deviation** , defined as the average distance of the observations from their mean:

$$s = \sqrt{s^2}$$

The standard deviation for cps is then:

$$s = \sqrt{2.90} = 1.70$$

Now, the calculation indicates that the standard deviation is 1.70 chirps per second, a more informative number to describe our data. The standard deviation is the most common measure of spread used in natural resources disciplines.

The variance (var()) and standard deviation (sd()) functions are calculated as:

```
var(chirps$cps)
```

```
## [1] 2.896952
```

```
sd(chirps$cps)
```

```
## [1] 1.702044
```

Another measure of spread is the **coefficient of variation (CV)**, a widely used statistic that "standardizes" the standard deviation relative to its mean. If we were only to look at the standard deviation of two different variables, their magnitudes might be vastly different. However, by standardizing the two variables we can compare how variable each is relative to its mean value. The CV is always expressed as a percent:

$$CV = \frac{s}{\bar{x}} \times 100$$

The CV for the chirps data then is:

$$CV = \frac{1.70}{16.65} \times 100 = 10.22\%$$

A CV of 10.33% is relatively low when investigating biological and environmental data. For comparison, most variables measured in a forest setting, such as tree diameters and volume, have CVs around 100% (Freese 1962). In R, the standard deviation and mean can be combined in a single line of code to calculate the CV:

```
(sd(chirps$cps) / mean(chirps$cps)) * 100
```

```
## [1] 10.22044
```

2.2.3   Measures of position

We are often interested to know how a particular value can be compared to other values in a data set. One value that uses the mean and standard deviation to measure the position of a variable of interest is the **z-score**. The z-score measures how different the data are from what we would expect (i.e., the mean). We'll revisit the z-score when we get into hypothesis testing, but a few important things to know include:

- a z-score equal to 0 signals that a value is equal to the mean,
- a z-score less than 0 signals that a value is less than the mean, and
- a z-score greater than 0 signals that a value is greater than the mean.

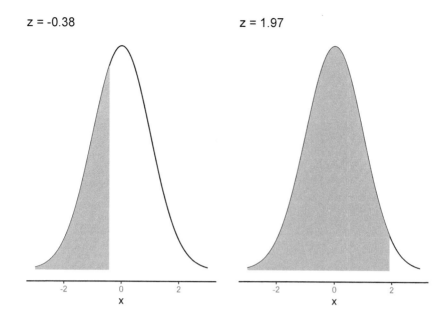

FIGURE 2.2 Z-scores for 16 and 20 chirps per second.

The value of the z-score tells you how many standard deviations a value x_i is away from the mean $\bar{x}$. Here, we will compare the value to the sample mean and standard deviation:

$$z_i = \frac{x_i - \bar{x}}{s}$$

In the chirps data, a z-score of 0 indicates that the value of interest x_i is equal to the mean. The figure below shows the corresponding z-scores if the number of chirps per second are 16 and 20, two values that are slightly lower and tremendously greater than the mean. Slightly less than half of observations would be expected to be less than 16. The majority of observations would be expected to be less than 20. In other words, 16 and 20 chirps per second are 0.38 and 1.97 standard deviations away from the mean, respectively.

We saw the advantages of a box plot from the last chapter and described how it shows the minimum, first quartile (Q1), median, third quartile (Q3), and maximum values of a variable. Here is a box plot for the chirps per second:

```
ggplot(data = chirps, aes(x = 1, y = cps)) +
  geom_boxplot() +
  scale_x_continuous(breaks = NULL) +
  labs(y = "Chirps per second")
```

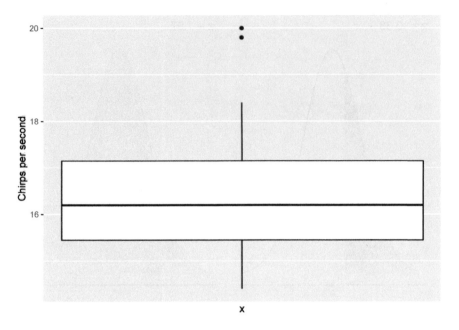

In the box plot, notice that there are two outliers that are shown as points. These correspond to the values at 19.8 and 20.0 chirps per second. By default, the whiskers using `geom_boxplot()` will extend no more than 1.5 times the **interquartile range (IQR)**, or the difference between Q3 and Q1 on either the lower or upper tail of the distribution. The IQR can be thought of as a range for the middle 50% of data. Hence, the IQR may be preferred over the range because it is less sensitive to extreme values.

The `coef` = statement can be added within `geom_boxplot()` to extend the whiskers a greater length from the IQR. For example, we can create a box plot that extends the whiskers to three times the IQR and will "cover up" the outliers:

```
ggplot(data = chirps, aes(x = 1, y = cps)) +
  geom_boxplot(coef = 3) +
  scale_x_continuous(breaks = NULL) +
  labs(y = "Chirps per second")
```

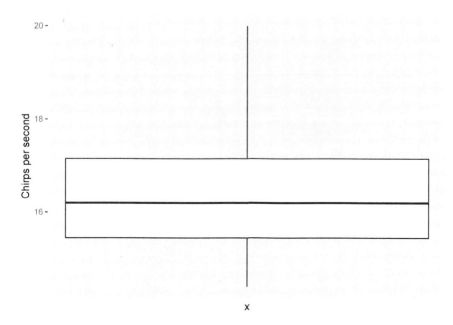

After viewing the box plot, we can also see that the data are right-skewed, because the difference between the maximum value and median is larger than the difference between the minimum value and median. More evidence of a right-skewed distribution is that the median (16.2) is less than the mean (16.65), indicating more large values and makes the mean larger.

DATA ANALYSIS TIP The classification of whether or not an observation is an outlier varies widely by discipline, the kind of instrumentation used, and other characteristics of the data. It is best to always check suspect observations with thorough analysis. Fortunately, after visualizing and summarizing data with descriptive statistics, you are in a great position to identify bad data before you begin your other statistical analyses.

2.2.4 Exercises

2.1 Ant species richness was sampled in bogs and forests at 22 sites in Connecticut, Massachusetts, and Vermont, USA (Gotelli and Ellison 2002). Environmental variables in the **ant** data set from the **stats4nr** package include

FIGURE 2.3 The following exercises use a dataset on ant species richness. Photo: Unitedbee Clicks on Unsplash.

species richness at each site by ecotype (forest or bog), latitude (lat), and elevation (elev). The data are contained in the **spprich** data set . Read in the data set and run the following code to create separate data frames for the two ecosystem types:

```
library(stats4nr)
head(ant)
```

```
## # A tibble: 6 x 5
##    site  ecotype spprich   lat  elev
##    <chr> <chr>     <dbl> <dbl> <dbl>
## 1 TPB   Forest        6  42.0   389
## 2 HBC   Forest       16  42       8
## 3 CKB   Forest       18  42.0   152
## 4 SKP   Forest       17  42.0     1
## 5 CB    Forest        9  42.0   210
## 6 RP    Forest       15  42.2    78
```

```
forest <- ant %>%
    filter(ecotype == "Forest")

bog <- ant %>%
    filter(ecotype == "Bog")
```

Use R functions to calculate the mean, median, and mode for the species richness in the forest and bog data sets. Which ecosystem type has a greater mean species richness?

2.2 Create a side-by-side violin plot that shows ant species richness on the x-axis with ecosystem type on the y-axis.

2.3 Learn about and use the `range()` and `IQR()` functions to calculate the range and interquartile range for species richness in the forest and bog data sets.

2.4 Calculate the standard deviation and coefficient variation for species richness in the forest and bog data sets. While both measurements reflect a measure of variability, your results will indicate a different answer to the question "Is ant species richness more variable in forests or bogs?" Explain why you obtain different results depending on whether you use the standard deviation or coefficient of variation.

2.5 How many standard deviations away from the mean are 7 species of ants? Write R code to make two calculations: one for the forest and one for the bog data set.

2.6 Instead of using the `filter()` function to create the **forest** and **bog** data sets as we did in exercise 2.1, we could also group the data by `ecotype` and summarize the data to report the summary statistics. Write R code using the **dplyr** functions `group_by()` and `summarize()` to create a summary of the data that shows the mean and standard deviation for the forest and bog ecosystems. HINT: See how we created the **elm_summ** data set in Chapter 1.

2.3 Random variables

Random variables are probability models that have a specific distribution. Any random variable will take on a numerical value that describes the outcomes of some chance process. Because of this, random variables are closely linked with probability distributions. Random variables can be either discrete or continuous.

TABLE 2.2 Probabilities of seeing a golden-winged warbler.

Number of warblers seen	Probability
0	0.6
1	0.1
4	0.3

2.3.1 Discrete random variables

A **discrete random variable** X has a countable number of possible values where one probability is associated with an individual value. In other words, if we can find a way to list all possible outcomes for a random variable and assign probabilities to each one, we have a discrete random variable.

Since discrete random variables can only take on a discrete number of values, there are two rules that apply to them. First, all probabilities take on a value between 0 and 1 (inclusive). Second, the sum of all probabilities must equal 1. Common discrete random variables used in natural resources include the binomial, geometric, Poisson, and negative binomial.

2.3.1.1 Mean and standard deviation of a discrete random variable

Consider as an example the following probability mass function that a birder has recorded after her research on the golden-winged warbler, a songbird that relies on open areas and patchy woodland habitat for nesting. The researcher noted that it was most likely to see no warblers during a research visit and never saw two or three warblers:

FIGURE 2.4 A golden-winged warbler. Photo: Natural Resources Conservation Service.

The warbler data can also be represented with the following probability histogram:

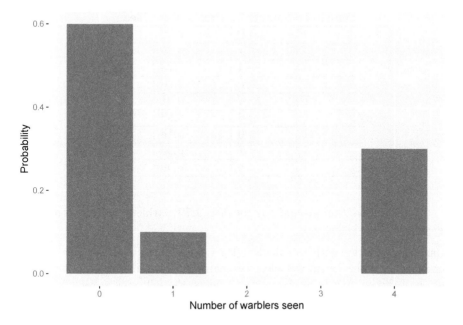

We can use the warbler data to find the mean and standard deviation of this discrete random variable. Let y denote the number of warblers seen and μ and σ^2 its mean and variance, respectively:

$$\mu = \sum yP(y)$$

$$\mu = (0 \times 0.6) + (1 \times 0.1) + (4 \times 0.3)$$

$$\mu = 1.3$$

So, on average we would expect to see 1.3 warblers given this discrete probability distribution. The variance and standard deviation can also be calculated:

$$\sigma^2 = \sum (y - \mu)^2 P(y)$$

$$\sigma^2 = (0 - 1.3)^2 \times (0.6) + (1 - 1.3)^2 \times (0.1) + (4 - 1.3)^2 \times (0.3)$$

$$\sigma^2 = 3.21$$

$$\sigma = \sqrt{\sigma^2} = 1.79$$

TABLE 2.3 Probabilities of seeing a golden-winged warbler with two samples.

Sample 1	Sample 2	Probability	Mean
0	0	0.36	0.0
0	1	0.06	0.5
0	4	0.18	2.0
1	0	0.06	0.5
1	1	0.01	1.0
1	4	0.03	2.5
4	0	0.18	2.0
4	1	0.03	2.5
4	4	0.09	4.0

So, a typical deviation around the mean is 1.79 warblers, as represented by the standard deviation σ.

Consider now if we wish to calculate the sample mean and variance for the number of warblers when we select two samples. To find the sampling distribution of the number of warblers, we could do the following steps:

- list all possible samples than could be observed and their probabilities,
- calculate the mean of each sample,
- list all unique values of the sample mean, and
- sum over all samples to calculate their probabilities.

These steps create a larger number of possibilities:

As you can see, even for a sample size of two, we're beginning to do calculations for a large number of scenarios. The warbler data with two samples can also be seen with the probability histogram:

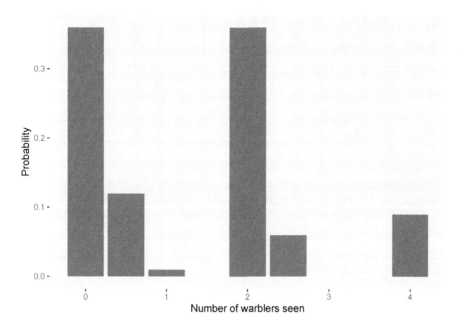

2.3.2 Sampling distribution of the mean

Note the differences between the probability histograms of one and two samples. We will denote the sample mean of collecting two samples as

$$\mu_{\bar{y}} = (0 \times 0.36) + (0.5 \times 0.12) + ... + (4 \times 0.09)$$

$$\mu_{\bar{y}} = 1.3$$

So, even after collecting two samples, the mean after collecting two samples is identical to the population mean. The variance and standard deviation of two samples can also be calculated:

$$\sigma_{\bar{y}}^2 = \sum (y - \mu)^2 P(y)$$

$$\sigma_{\bar{y}}^2 = (0 - 1.3)^2 \times (0.36) + (0.5 - 1.3)^2 \times (0.12) + ... + (4 - 1.3)^2 \times (0.09)$$

$$\sigma_{\bar{y}}^2 = 1.61$$

$$\sigma_{\bar{y}} = \sqrt{\sigma_{\bar{y}}^2} = \sqrt{1.61} = 1.27$$

Notice the standard deviation of $\bar{y}$ is smaller than the standard deviation for σ. We call the value 1.27 the **standard error**, defined as the standard deviation of a sampling distribution. The standard error can also be calculated by dividing the standard deviation by the square root of the sample size:

$$\sigma_{\bar{y}} = \frac{\sigma}{\sqrt{n}}$$

$$\sigma_{\bar{y}} = \frac{1.79}{\sqrt{2}} = 1.27$$

What would happen if we continued to draw more samples of the warbler sightings, say by taking three, five, 10, and 15 samples? No doubt our calculations would become more complex. We are coming up to a big idea in statistics.

2.3.2.1   The central limit theorem

We can continue to take all possible samples of size n and calculate the sample mean and standard error of the sampling distribution. Using the warbler data, here is the sampling distribution as we increase from three to 15 samples:

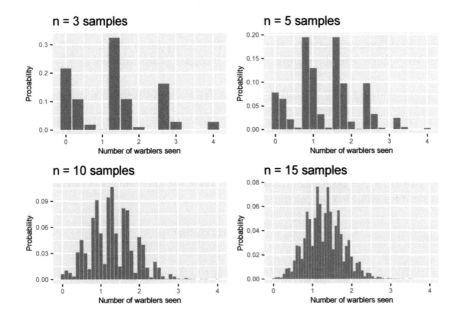

FIGURE 2.5 Visualizing the central limit theorem, from three to 15 samples.

Note that the graphs of the sampling distributions change in appearance as you increase the number of samples. This is the **central limit theorem** in action. The central limit theorem shows that if you take a large number of samples, the distribution of means will approximate a normal distribution, or a bell-shaped curve. This is a remarkable finding because most population distributions are not normal (i.e., the warbler observations). When a sample is large enough (generally at least 25 observations in the natural resources), the distribution of sample means is very close to normal.

2.3.3   Continuous random variables

A **continuous random variable** has an "uncountable" number of values where the probability of occurrences move continuously from one value to the next. A continuous random variable Y can be described by a probability density function $f(y)$, which can be thought of as the area beneath a curve.

For any continuous random variable, the total area beneath the curve must sum to one. Hence, $P(a < y < b)$ can be considered the area beneath the curve between a and b. Common continuous random variables used in natural resources include the normal, Chi-square, uniform, exponential, and Weibull distributions.

2.3.3.1   Mean and standard deviation of a continuous random variable

When we found the mean and standard deviation for the warbler data, a discrete random variable, we summed across values. However, because continuous random variables have an "uncountable" number of values, the right approach is to integrate across all values to find its mean and standard deviation.

The **normal distribution** is the foundation for statistical inference throughout this book. The normal distribution describes many kinds of data. Numerous statistical tests assume that data are normally distributed. The mean and standard deviation of a normal distribution are denoted by μ and σ, respectively.

The mean of a normal distribution is the center of a symmetric bell-shaped curve. Its standard deviation is the distance from the center to specified points on either side. A special case of the normal distribution with a mean of 0 and standard deviation of 1, or N(0,1) is termed the **standard normal distribution**. Consider, for example, two distributions that have the same mean ($\mu = 42$) but different standard deviations ($\sigma = 3$) and ($\sigma = 8$):

The distributions show that the N(42, 3) distribution is much narrower than the N(42, 8) distribution. This is important because the mean and standard deviation of a normal distribution help define the **empirical rule**, a rule

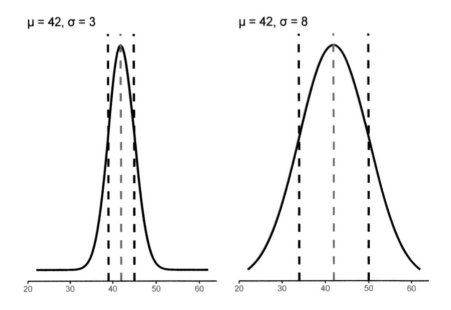

FIGURE 2.6 The normal distribution with the same mean and different standard deviations.

that describes the approximate percentages of the range of observations. The empirical rule, also referred to as the 68-95-99.7% rule, states that:

- approximately 68% of the observations fall within σ of μ,
- approximately 95% of the observations fall within 2σ of μ, and
- approximately 99.7% of the observations fall within 3σ of μ.

We can calculate a z-score for any normal distribution. For example, for the $N(42, 3)$ distribution, we can calculate the z-score for the value 38:

$$z_{38} = \frac{38 - 42}{3} = -1.33$$

In other words, the value 38 is 1.33 standard deviations less than the mean value of 42. In R, we can write a function called z_score() that calculates a z-score using the three values: the observation of interest x_i, the population mean μ, and the population standard deviation σ:

```
z_score <- function(x_i, pop_mean, pop_sd){
    z <- (x_i - pop_mean) / pop_sd
```

```
    return(z)
}
```

Then, we can add the values of interest to use the R function to calculate the z-score:

```
z_score(x_i = 38, pop_mean = 42, pop_sd = 3)
```

```
## [1] -1.333333
```

So, we know that 38 is 1.33 standard deviations less than the mean.

2.3.4   Exercises

2.7 A deer hunter tells you that there is a 0.7 probability of seeing no deer when he goes hunting, a 0.1 probability of seeing one deer when he goes hunting, and a 0.2 probability of seeing two deer when he goes hunting. Use calculations in R to determine the mean μ and standard deviation σ of the number of deer seen from this discrete random variable.

2.8 Mech (2006) published results from a study on the age structure of a population of wolves in the Superior National Forest of northeastern Minnesota, USA. Results indicated a high natural population turnover in wolves.

Looking at Table 1 in the Mech (2006) paper, we can enter the ages and number of wolves in the study using the tribble() function. After creating the data set, plot the wolf age data using a bar plot.

```
wolves <- tribble(
  ~age, ~num_wolves,
  0, 10,
  1, 9,
  2, 21,
  3, 8,
  4, 9,
  5, 5,
  6, 2,
  7, 2,
  8, 1,
```

```
 9, 2
)
```

2.9 In the wolf data, use calculations in R to determine the mean μ and standard deviation σ of the wolf ages.

2.10 Use the `z_score()` function presented earlier in the chapter to calculate the z-score for a wolf that is six years old.

2.11 Historical data for the amount of lead in drinking water suggests its mean value is $\mu = 6.3$ parts per billion (ppb) with a standard deviation of 0.8 parts per billion. Using concepts from the empirical rule, use R to calculate the upper and lower bounds of the values that are one, two, and three standard deviations away from the mean value.

2.12 An environmental specialist collects data on 15 drinking water samples and calculated a mean lead content of 8.0 ppb. Using the population data in the previous problem, calculate the z-score for this sample.

2.4   Summary

Descriptive statistics are essential components that allow you to better understand the characteristics of your data. Measures of central tendency, spread, and position are values that describe your data and help place it in the context of other observations and data.

By learning about random variables, we have introduced the concepts of probability and how they lead to different outcomes of a chance process. Discrete random variables have a countable number of outcomes whereas continuous random variables have an infinite number of possibilities along a continuous scale. We were introduced to the normal distribution, which can be described by a mean and standard deviation and will form the basis of statistical inference. The concepts we learned about the probability as it relates to the normal distribution will be revisited in subsequent chapters when we discuss statistical inference and hypothesis testing.

2.5 References

Freese, F. 1962. *Elementary forest sampling.* USDA Forest Service Agricultural Handbook No. 232. 91 p.

Gotelli, N.J., and A.M. Ellison. 2002. Biogeography at a regional scale: determinants of ant species density in New England bogs and forests. *Ecology* 83: 1604–1609.

Mech, L.D. 2006. Estimated age structure of wolves in northeastern Minnesota. *Journal of Wildlife Management* 70: 1481–1483.

Pierce, G.W. 1943. The songs of insects. *Journal of the Franklin Institute* 236: 141–146.

Poncet, P. 2019. modeest: mode estimation. R package version 2.4.0. Available at: https://CRAN.R-project.org/package=modeest

3

Probability

3.1 Introduction

Probability is so widespread in statistics that we've already talked about it in depth in the first two chapters. Probability is based on **randomness**, the process by which outcomes are uncertain with some distribution. We may or may not know what the distribution of outcomes is, but can find it after a large number of repetitions. By understanding randomness and probability, we will ease into the discipline of statistical inference, where we will talk about hypothesis testing and estimation.

Consider a tree that is subject to infestation by a non-native insect that can lead to its death. The concepts of probability are based on several definitions:

- A **random experiment** is any observation or measurement that has an uncertain outcome. For example, a forest health specialist may conduct an assessment of the tree to determine whether or not the insect is present (i.e., the tree shows symptoms of the insect damage) or absent (the tree is healthy).

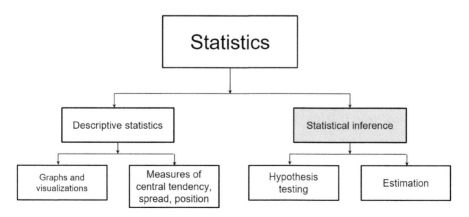

FIGURE 3.1 This chapter focuses on probability, and introduces topics related to statistical inference.

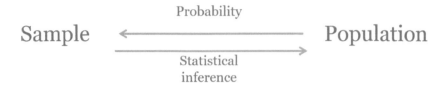

FIGURE 3.2 Probability allows us to use information from populations to inform samples.

- A **sample space** is a list of all possible outcomes of an experiment. For example, the insect is present on the tree or not.
- An **event** is any collection of one or more outcomes from an experiment. For example, the forest health specialist visits one tree or one hundred trees and records whether or not the insect is present.
- The **probability** of any outcome is the proportion of times an event would occur over repeated observations.

Probability allows us to use information from populations to inform samples. Statistical inference allows us to generalize from samples to make statements about populations. Together, these relationships will serve as the foundation for much of the rest of this book.

Note the differences between the statements below with regard to probability versus statistical inference. Are they the same?

A probability approach: A forest health specialist manages 100 ash trees in a city. Fifteen of these trees show symptoms of emerald ash borer, a invasive insect that leads to ash tree mortality in North America. The forest health specialist samples 10 of these ash trees. What is the chance that three of the ten trees show symptoms of emerald ash borer?

A statistical inference approach: Ten ash trees are sampled in a metropolitan area with 1000 ash trees. Three ash trees show symptoms of emerald ash borer. What is the proportion of ash trees that have emerald ash borer?

This chapter will discuss how the concepts of probability can be used in R to better understand natural resources data. This will allow us to make inferences that will aid us in making statements about data.

3.2 The five rules of probability

The following rules introduce guidelines for probability. These will assist us when performing calculations and creating visualizations of events.

3.2.1 Rule 1: Probabilities are between 0 and 1

Any probability is a value between 0 and 1. Values range from 0 (impossible) to 1 (certain). You may often hear about probabilities discussed in terms of percentages, likelihoods, or odds. The key when using probabilities in calculations is that they are always expressed in decimal form, and not percentages. For example, there is a 0.60 probability of having a drought this year, not a 60% chance of having a drought.

Understanding a probability value is important, but where do probabilities come from? If we know how many outcomes to expect, this is an example of using a **classical approach** to assign probabilities. If an experiment has n outcomes, we can assign a probability of $1/n$ to each outcome. For example, what is the probability we flip a coin and get a head?

We could also rely on past experiments or trends to assign probabilities. In this example, probabilities are derived from **historical data** or experiments. Due to the experimental nature of this approach, it is important to note that different experiments will typically result in different probabilities. For example, if you flip a coin ten times you might get a head four times. If you repeat the experiment, you might get a head seven times. In other words, if n_A is the number of times that event A occurred, n_A/n represents the frequency at which A occurs.

A third way to assign probabilities is using a **subjective approach**. In this approach, the probability is based on intuition or prior knowledge. While subjective approaches may be useful, they are not always based on data or can make important assumptions about the variables of interest. For example, a researcher might say there is a 0.60 probability of having a drought this year based on her "gut instinct."

DATA ANALYSIS TIP: Many disciplines, such as sports and weather forecasting, use a mix of approaches to determine probabilities. *The Old Farmer's Almanac*, a reference book for

TABLE 3.1 Common trees in North America.

Common name	Scientific name	Clade	Deciduous
Sugar maple	*Acer saccharum*	Angiosperm	Yes
Yellow poplar	*Liriodendron tulipifera*	Angiosperm	Yes
White oak	*Quercus alba*	Angiosperm	Yes
Red maple	*Acer rubrum*	Angiosperm	Yes
Northern red oak	*Quercus rubra*	Angiosperm	Yes
Douglas-fir	*Pseudotsuga menziesii*	Gymnosperm	No
Loblolly pine	*Pinus taeda*	Gymnosperm	No
Ponderosa pine	*Pinus ponderosa*	Gymnosperm	No
Lodgepole pine	*Pinus contorta*	Gymnosperm	No
Western hemlock	*Tsuga heterophylla*	Gymnosperm	No
Tamarack	*Larix laricina*	Gymnosperm	Yes
Bald cypress	*Taxodium distichum*	Gymnosperm	Yes

weather forecasts in the United States, indicates an average accuracy rate of 80% for its annual weather predictions. This rate is calculated by comparing their predicted changes in temperature and precipitation to the actual observed changes in different areas of the US (*The Old Farmer's Almanac* 2020). Note that these accuracy numbers are self-reported.

3.2.2 Rule 2: All probabilities sum to one

To determine if a probability model is appropriate, sum up all of the probabilities. Their sum should add to one. In other words, all probabilities of events in the sample space should add to one. Venn diagrams are useful to see how the events are related, such as the probability of events A and B occurring. A Venn diagram will show if two events are **mutually exclusive**, i.e., if two events can occur with one repetition of an experiment.

Consider the following example of common North American tree species and their clade (angiosperm or gymnosperm) and whether or not they are deciduous (i.e., their leaves fall after the growing season):

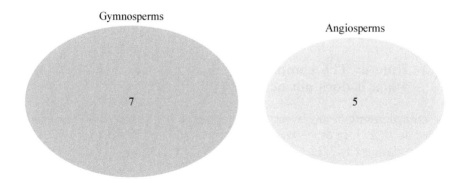

FIGURE 3.3 Venn diagram showing the clades of trees, events that are mutually exclusive.

We can say that the clade of a tree is mutually exclusive (or "disjoint") because a tree can only be a gymnosperm or angiosperm:

Venn diagrams will be useful to create and examine to make sure your model follows the rules of probability.

3.2.3 Rule 3: If two events are mutually exclusive, apply the addition rule

If two events do not share the same outcomes and are mutually exclusive, the probability that one or the other occurs is the sum of their individual probabilities. This is termed the **addition rule of probability**. This can be written as:

$$P(A \bigcup B) = P(A) + P(B)$$

The notation $P(A \bigcup B)$ can be thought of as the probability of A or B occurring. Take for example the **tree** data set with 10 different species. We can determine the probability of selecting a tree from the maple (*Acer*) or pine genus (*Pinus*). We know that $P(maple) = 2/10$ and $P(pine) = 3/10$. We can then write

$$P(\text{maple} \bigcup \text{pine}) = 2/12 + 3/12 = 0.4167$$

Maples and pines are common in the data set and there is about a 42% chance we would select one of those species at random.

3.2.4 Rule 4: The complement of a probability is measured when it does not occur

The **complement** of any event A is the probability that A does not occur, defined as $P(A^C)$. In other words, if the probability of choosing a maple tree from our data set is $2/12 = 0.1667$, the probability of not choosing a maple tree is $10/12 = 0.8333$. More generally, the complement of a probability can be written as $P(A^C) = 1 - P(A)$.

The idea of the complement of a probability is highlighted with a well-known observation in probability known as the birthday problem (Feller 1968). Suppose there are n people in a room at one time. What is the probability that two people in that room share the same birthday? Rather than finding the probability of an event, it's often easier to find the probability of it not occurring (i.e., the complement).

To determine the probability of a birthday match, we'll make a few assumptions:

- Birthdays are independent of one another. That is, there are no twins in the room.
- Birthdays are distributed evenly across the year.
- We'll consider only non-leap years with 365 days in the year.

Then we can say that the probability of two students not sharing the same birthday is $364/365$. If there are n people in the room, we can state that

$$P(\text{at least one birthday match}) = 1 - (\frac{364}{365} \times \frac{363}{365} \times ... \times \frac{365 - n}{365})$$

It turns out that if there are 23 people in a room, there is about an even chance that at least two people in the room will share the same birthday. These somewhat surprising results are a reflection that there are so many possible pairs of birthdays.

While it may be difficult to find the probability of an event, the birthday problem shows us that calculating the complement of an event is often easier.

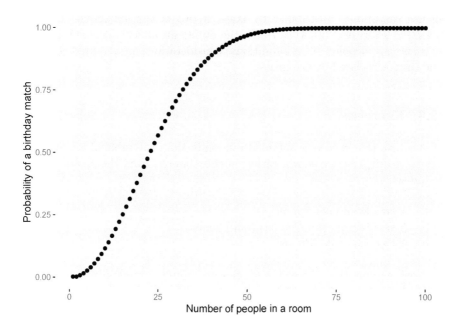

FIGURE 3.4 Probability of a birthday match with different number of people in a room.

3.2.5 Rule 5: If one event happening does not influence another, the events are independent

Two events are **independent** if knowing that one event occurs does not change the probability that the other occurs. In other words, if two events A and B do not influence each other, and if knowledge about one does not change the probability of the other, the events are independent.

The probability we assign to an event can change if we know that some other event has occurred. This is common in natural resources because we often collect data across time (e.g., weather patterns today can influence how plants grow tomorrow) and space (e.g., plants will not have the same growth rate if grown at different densities with varying levels of light and nutrients). When we calculate the probability that one event will happen given that another event has occurred, we are determining a **conditional probability**. The conditional probability of event A occurring given event B has already occurred can be written as:

$$P(A|B) = \frac{P(A \text{ and } B)}{P(B)}$$

We can interpret the pipe "|" by saying "given" or "conditioned on." Whether or not events are independent can inform how to calculate probabilities. You can use conditional probability to assess if two events are independent. If $P(A|B) = P(A)$ or $P(B|A) = P(B)$, the events are independent. Otherwise, the events are dependent and additional considerations are required to determine their associated probabilities.

3.2.6 Exercises

3.1 A single tree species is sampled randomly from the table of 12 common trees in North America shown earlier in this chapter. Find the probability of the following:

    a. A tree is an angiosperm.
    b. A tree is from the *Quercus* (oak) genus.
    c. A tree is not deciduous.

3.2 Now, assume that three tree species are sampled randomly (without replacement) from the table of 12 common trees in North America shown earlier in this chapter. Find the probability of the following:

    a. Three trees are sampled from the *Pinus* (pine) genus.
    b. Three trees sampled are gymnosperms.
    c. The first two trees sampled are gymnosperms and the last species is an angiosperm.
    d. Three trees sampled are angiosperms or three trees are sampled from the *Pinus* (pine) genus.

3.3 In a forest, 30% of the trees suffered damage from an insect and were affected by a disease. A total of 65% of those trees only suffered damage from an insect. What is the probability of a tree being affected by a disease given it already suffered damage from an insect?

3.4 Use the following function, `birthday()`, in R to calculate the probability of two people sharing the same birthday if 5, 20, and 40 people are in a room together (i.e., $n = 5$, 20, and 40). Note that q calculates each of the probabilities of not sharing the same birthday and p sums those probabilities. The probability is printed in the output.

```
birthday <- function(n){
  q <- 1 - (1:(n - 1)) / 365
  print(p <- 1 - prod(q))
}
```

3.5 By memory, list the birthdays and names for all of your close friends and family members. List any number of birthdays that come to memory. Use the birthday() function in the above problem to determine the probability of two of them sharing the same birthday. Do any of your close friends and family members share the same birthday?

BAYES' THEOREM: If prior knowledge related to an event is known, Bayes' Theorem can be used to determine the probability of an event. The probability of event A occurring given that B is true can be determined with the formula below. Bayes' Theorum is used so widely that it led to the creation of a discipline within statistics termed Bayesian statistics. More generally, the conditional probability $P(B/A)$ represents the likelihood of the event occurring, i.e., the degree to which event B is supported by event A. The prior probability is represented by $P(A)$, representing a subjective or objective belief about the probability before data are considered. The data themselves, also known as the "evidence," are represented by $P(B)$. The conditional probability $P(A/B)$ of interest is referred to as the posterior probability with Bayesian statistics.

$$P(A|B) = \frac{P(B|A)P(A)}{P(B)}$$

3.3   Probability functions in R

In R, we can use the sample() function to randomly sample from a list. This function will be helpful in many aspects of probability. For example, we can sample two random species from the **tree** data set:

```
sample(tree$`Common name`, 2)
```

```
## [1] "Lodgepole pine" "Yellow poplar"
```

By default, the sample() function will select from a list without replacement. We can add to the code to select two trees **with replacement**.

```
sample(tree$`Common name`, 2, replace = TRUE)
```

```
## [1] "Lodgepole pine"  "Western hemlock"
```

The prod() function multiplies several values together. This can be useful to when sampling without replacement, as the number of remaining samples will decrease by one after each iteration. For example, consider that we want to sample two species from the **tree** data set containing 12 species. We could find the probability of selecting two specific species as:

```
(1/12) * (1/11)
```

```
## [1] 0.007575758
```

Or with the prod() function we could write

```
1 / prod(12:11)
```

```
## [1] 0.007575758
```

where a colon : indicates a sequence of values to multiply, i.e., "from 12 to 11." The sum() function works similarly and will add a sequence of numbers together. The cumprod() and cumsum() functions provide the cumulative product and sum of a sequence of values.

When calculating probabilities in R, the choose() function calculates the total number of possibilities of a subset of events occurring. In our example using the **tree** data set, we can find the total number of possibilities for selecting two unique species from the larger list of 12 species:

```
choose(12, 2)
```

```
## [1] 66
```

There are 66 total possibilities for selecting two unique species from the list of 12 species. The choose() function represents a combination of outcomes and can be written as $\binom{n}{k}$, where n items are sampled k times without replacement. It can be calculated using factorials, i.e.,

$$\binom{n}{k} = \frac{n!}{k!(n-k)!}$$

In the **tree** data set, 12! (pronounced "12 factorial") equates to $12*11*10*...*1$. Hence,

$$\frac{12!}{2!(12-2)!}$$

results in the 66 total possibilities.

3.3.1 Exercises

3.6 A plant ecologist wants to randomly sample 5 plants from a total of 20 plants (numbered 1 through 20) to determine the presence of a foliage disease. Use the sample() function in R to determine which plants to sample without replacement.

3.7 Now, use the sample() function in R to sample 5 plants **with replacement**.

3.8 Use the choose() function to calculate the total number of possibilities of selecting four unique plants to measure from the 20 total plants.

3.9 Use the choose() function to calculate the probability of sampling (without replacement) the following four plant numbers: 6, 14, 15, and 19.

3.10 Instead of the choose() function, use the formula for the combination of outcomes to arrive at the same answer as in 3.9 but only by using the prod() function.

3.4 Discrete probabilities: the Bernoulli and binomial distributions

The Bernoulli and binomial distributions are two examples of discrete distributions. Understanding them can help to interpret probabilities. The **Bernoulli**

distribution is a discrete distribution having two possible outcomes: success or failure. A **binomial distribution** is a sum of Bernoulli trials. When the same chance process is repeated several times, we are often interested in whether a particular outcome does or does not happen on each repetition. In some cases, the number of repeated trials is fixed in advance, and we are interested in the number of times a particular event occurs.

Flipping a coin and observing whether it is heads or tails is an example of a Bernoulli trial. With a fair coin, the probability of getting a heads is $p = 0.5$. The Bernoulli distribution is the simplest discrete distribution, and is the foundation for other more complicated discrete distributions.

In mathematical terms, Binomial = Bernoulli + Bernoulli + ... + Bernoulli. The binomial distribution is measured by the count of successes (X) with parameters n and p, where:

- n is the number of trials of the chance process,
- p is the probability of a success on any one trial, and
- the possible values of X are the whole numbers from 0 to n.

If a count has a binomial distribution with n trials and probability of success p, the mean can be determined as np and the standard deviation as $np(1-p)$.

The formula for finding the probability of exactly x successes is:

$$P(X = x) = \frac{n!}{x!(n-x)!}p^x(1-p)^{n-x}$$

The binomial distribution requires knowledge of the combination of outcomes to determine the probability of x successes.

As an example application of the binomial distribution, consider the relationship between ruffed grouse (*Bonasa umbellus*) and West Nile virus (from the genus *Flavivirus*). Wildlife biologists are interested in knowing how the virus, a mosquito-borne disease, impacts the health of ruffed grouse, a forest bird located across North America and a popular game species. Researchers have hypothesized that the introduction of West Nile virus in the early 2000s in North America has led to population declines in ruffed grouse (Stauffer et al. 2017).

In 2018, a multi-state effort examined the presence of West Nile virus in ruffed grouse populations across the US Lake States. The study concluded that while the West Nile virus was present in the region, grouse that were exposed to West Nile did not always die and many survived. The total number of ruffed grouse sampled in each of the three states, and those that showed antibodies consistent with West Nile virus is shown in the table below:

FIGURE 3.5 Ruffed grouse, a North American forest bird. Image: Ruffed Grouse Society.

Using the grouse data, we can think of whether or not a grouse has West Nile as a Bernoulli trial. If a group of grouse are sampled, some number of them may have the West Nile virus, and we have a binomial distribution.

As an example, consider a random sample of six grouse from Michigan. What is the probability that none of these grouse will have the West Nile virus? The problem can be solved with the binomial formula assuming $x = 0$ grouse are infected with West Nile from a total of $n = 6$ grouse. The probability of a grouse having West Nile virus in Michigan is $p = 0.13$:

$$P(X = 0) = \frac{6!}{0!(6-0)!}0.13^0(1 - 0.13)^{6-0} = 0.4336$$

Or, there would be about a 43% chance that if you sampled six grouse from Michigan, none of them would have the West Nile virus.

Below shows a function p.binomial() that creates "pin probabilities" showing the results of a binomial problem. The visualization shows a sample of $n = 6$ ruffed grouse from Michigan and calculates the probability of each possible

TABLE 3.2 State results of West Nile occurrence in ruffed grouse.

State	Number of grouse sampled	Number of grouse with West Nile	Proportion of grouse with West Nile
Michigan	213	28	0.13
Minnesota	273	34	0.12
Wisconsin	235	68	0.29

number of grouse being infected with West Nile. Note the ymax and xmax
statements specify the length of the y and x axes, respectively:

```r
p.binomial <- function(n, p, title, ymax, xmax){
  x <- dbinom(0:n, size = n, prob = p)
  barplot(x, ylim = c(0, ymax), xlim = c(0, xmax),
          names.arg = 0:n,
          xlab = "Number of grouse with West Nile",
          ylab = "Probability",
          main = sprintf(paste(title)))
}

p.binomial(6, 0.13,
           "Binomial distribution (n = 6, p = 0.13)",
           0.5, 7)
```

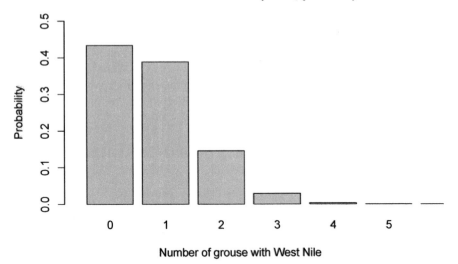

We can see that if six grouse are sampled in Michigan, the most probable
outcome is that zero ruffed grouse will have West Nile. We can modify the
function to visualize how many grouse in Wisconsin may be infected with
West Nile virus if a sample of $n = 50$ are taken:

```
p.binomial(50, 0.29, "Binomial distribution (n = 50, p = 0.29)",
           0.15, 51)
```

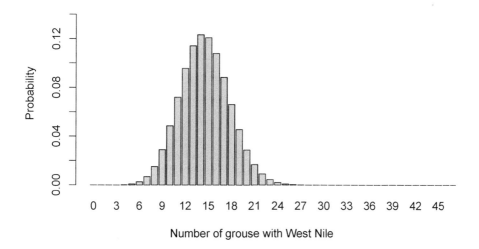

We can see that if 50 grouse are sampled in Wisconsin, the most probable outcome is that between 12 and 16 ruffed grouse will have West Nile.

One of the key characteristics of a discrete distribution is that values may only take on distinct values. In R, several density functions allow for calculating probabilities of events from specific distributions. The dbinom() function provides the density for a binomial distribution with parameters x, size, and prob. We can use the dbinom() function to find the probability that no grouse from a random sample of six grouse in Michigan will have the West Nile virus:

```
dbinom(x = 0, size = 6, prob = 0.13)
```

```
## [1] 0.4336262
```

We might be interested to know the probability of 0 or 1 or 2 grouse being infected with West Nile from a random sample of six grouse in Michigan. In this case, we can add a series of binomial densities together to determine the probability:

```
dbinom(x = 0, size = 6, prob = 0.13) +
  dbinom(x = 1, size = 6, prob = 0.13) +
  dbinom(x = 2, size = 6, prob = 0.13)
```

[1] 0.9676241

Alternatively, the pbinom() function provides the distribution function of the binomial. Probabilities are summed up to and including the value specified in the q parameter:

```
pbinom(q = 2, size = 6, prob = 0.13)
```

[1] 0.9676241

While the binomial distribution is common in natural resources, other discrete random variables include the geometric, Poisson, and negative binomial.

3.4.1   Exercises

3.11 Consider a random sample of 25 grouse. Use the table that shows the state results of West Nile occurrence and the binomial functions to calculate the following probabilities.

  a. Use the p.binomial() function to make a plot of the probabilities of grouse in Minnesota being infected with West Nile virus.
  b. What is the probability that five grouse from Minnesota will have the virus?
  c. What is the probability that five or fewer grouse from Minnesota will have the virus?
  d. What is the probability that six or more grouse from Minnesota will have the virus?

3.12 You are tasked with inspecting the facilities of companies that store hazardous waste as a part of their operations. If the company is not storing the waste safely (e.g., the material is not labeled properly or it is improperly stored), you will issue the company a violation. You inspect eight facilities for hazardous waste compliance and you know from historical records that there is a probability of 0.24 they will be issued a violation.

a. Modify the p.binomial() code to make a plot of the probabilities of issuing violations.
b. Use R code to calculate the mean and standard deviation of violations for this binomial distribution.
c. What is the probability that two facilities will be issued a violation?
d. What is the probability that two or fewer facilities will be issued a violation?

3.5 Continuous probabilities: the normal distribution

In Chapter 2, we were introduced to the normal distribution, one of the most widely used continuous distributions in statistics. We looked at an example of an N(42, 3) distribution, or a normally distributed population with a mean of 42 and a standard deviation of 3. We used the z-score calculation to determine that a value of interest, 38, was 1.33 standard deviations less than the mean. Similar to the dbinom() function, we can use the dnorm() function to find the density of the normal distribution given its mean and standard deviation:

```
dnorm(x = 38, mean = 42, sd = 3)
```

```
## [1] 0.05467002
```

The pnorm() function finds the cumulative values for a normal distribution depending on the specified parameters.

```
pnorm(q = 38, mean = 42, sd = 3)
```

```
## [1] 0.09121122
```

In other words, for an N(42, 3) distribution we can expect approximately 9% of observations to fall below 38. In contrast, its complement (or 91% of all observations) would be expected to be found greater than 38. This becomes more apparent after we plot it on the normal distribution:

z = -1.33

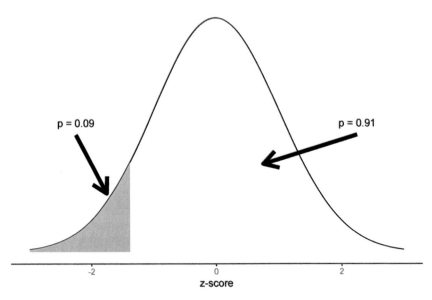

To find the quantile function or a random number for the normal distribution, the qnorm() and rnorm() functions can be used. The same d, p, q, and r prefixes can be used for many other kinds of discrete and continuous distributions in R. For example, the dexp(), pexp(), qexp(), and rexp() functions can be used to determine the density, distribution, quantile, and random number for the exponential distribution.

In addition to the normal distribution, the cumulative distribution function will be valuable for calculating probabilities other continuous distributions as well. These include the Chi-square, uniform, exponential, and Weibull distributions.

3.5.1   Exercises

3.13 Historical data for the amount of lead in drinking water suggests its mean value is $\mu = 6.3$ parts per billion (ppb) with a standard deviation of 0.8 parts per billion.

   a. An environmental specialist collects data on 15 drinking water samples and calculates a mean lead content of 8.0 ppb. Is the mean of the specialist's data unusually large?
   b. Use the rnorm() function to create a series of 100 random numbers sampled from the drinking water data.

c. Make a histogram of the series of 100 random numbers using gg-plot() and describe the pattern you see.

3.14 Data from the Minnesota Department of Natural Resources in 2017 suggest the average mid-winter pack size of gray wolves (*Canis lupus*) has a mean of $\mu = 4.8$ wolves with a standard deviation of $\sigma = 1.5$ wolves. Use this information to answer the following questions.

a. Assuming the pack sizes of wolves follows a normal distribution, calculate the probability of observing a pack size of two wolves. Would seeing this be a rare event?
b. Calculate the probability of observing a wolf pack size between four and seven wolves.

3.6 Summary

Probabilities are so widespread that they come up in nearly every topic related to statistics. The five rules of probability can be used in future statistical calculations. While we focused extensively on the binomial and normal distribution in this chapter, there any many other kinds of discrete and continuous distributions that we can apply probabilities to. In the topic of hypothesis testing we will discuss p-values, or probability values, measures that assess the probability of obtaining a result under specific conditions. Having a solid grasp of probability is required as we continue applying these concepts of statistical inference.

3.7 References

Feller, W. 1968. *An introduction to probability theory and its applications (Vol. 1)*. New York: Wiley. 509 p.

The old farmer's almanac. 2020. How accurate is the Old Farmer's Almanac weather forecast? Looking back on our winter 2019-2020 forecast, Available at: https://www.almanac.com/how-accurate-old-farmers-almanacs-weather-forecast

Stauffer, G.E., D.A. Miller, L.M. Williams, J. Brown. 2018. Ruffed grouse population declines after introduction of West Nile virus. *Journal of Wildlife Management* 82: 165–172.

4

Hypothesis tests for means and variances

4.1 Introduction

This chapter will introduce the concepts of **hypothesis tests**. At their simplest, hypothesis tests are claims that we make about phenomena. Hypothesis tests are tools that allow us to make statements about whether or not the means or variances of data are equal to a unique value or whether or not two groups of data are different.

Now that we have a foundation of probability, hypothesis tests will make up a large component of how we use statistical tests to make inference about populations:

Recall that we use different notation to denote what we mean by the sample and the population. This is essential in hypothesis testing because we collect data from samples to inform differences across populations:

- For counts, the population parameter is represented by N and the sample statistic by n.

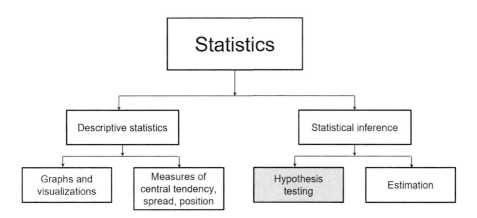

FIGURE 4.1 The next chapters focus on hypothesis tests, a component of statistical inference.

- For the mean, the population parameter is represented by μ and the sample statistic by $\mu_{\bar{y}}$ or $\bar{y}$.
- For the variance, the population parameter is represented by σ^2 and the sample statistic by s^2.
- For the standard deviation, the population parameter is represented by σ and the sample statistic by s.

This chapter will provide an introduction to the theory and concepts behind hypothesis testing and will introduce two new distributions to compare the mean and variance of a variable of interest.

4.2 Statistical significance and hypothesis tests

A test of statistical significance is a formal procedure for comparing observed data with a claim whose truth we want to assess. For example, consider a sample that determined the average size of a pack of wolves to be 4.8 wolves. A wildlife biologist is skeptical of this finding, claiming that she believes the average wolf pack size is seven wolves. Is the mean value of 4.8 significantly different from 7 wolves? Consider if wolf pack sizes in a different area were sampled with a mean size of 3.5 wolves. Are 4.8 and 3.5 wolves significantly different from each other?

A test of statistical significance evaluates a statement about a parameter such as the population mean or variance σ^2. The results of a significance test are expressed in terms of a probability termed the p-value. A **p-value** is the probability under a specified statistical model that a statistical summary of the data would be equal to or more extreme than its observed value. P-values do not measure the probability that a hypothesis is true nor should scientific conclusions or decisions be made if the p-values fall below or above some threshold value (Wasserstein and Lazar 2016). We will use statistical tests and the p-values that result from them as another metric when we perform hypothesis tests.

Any significance test begins by stating the claims we want to compare. These claims are termed the null and alternative hypotheses. The **null hypothesis**, denoted H_0, is the hypothesis being tested. The null hypothesis is assumed to be true, but your data may provide evidence against it. The **alternative hypothesis**, denoted H_A is the hypothesis that may be supported based on your data. Every hypothesis test results in one of two outcomes:

- The null hypothesis is rejected (i.e., the data support the alternative).
- The null hypothesis is accepted, or we fail to reject the null hypothesis (i.e., the data support the null).

When we draw a conclusion from a test of significance (assuming our null hypothesis is true), we hope it will be correct. However, due to chance variation, our conclusions may sometimes be incorrect. Type I error and Type II error are two ways that measure the error associated with hypothesis tests.

- **Type I error** occurs when we reject the null hypothesis when it is in fact true. Type I error reflects the **level of significance** of a hypothesis test, denoted by α.
- **Type II error** occurs when we fail to reject the null hypothesis when it is in fact false, denoted by β.

The complements of both Type I and Type II errors represent the correct decisions of a hypothesis test. The probability of correctly rejecting the null hypothesis is when it is indeed false is $1 - \beta$, also known as the **statistical power** of a hypothesis test.

4.2.1 Setting up hypothesis tests

Hypothesis tests can be one- or two-tailed. This choice depends on how the null hypotheses are set up. For example, consider a one-sample t-test for a mean, which will be described more in depth later in this chapter. This test compares the difference between a mean value μ and a hypothesized value in the population. This hypothesized mean is often a unique or interesting value that we wish to compare to, denoted μ_0.

Our null hypothesis can be written as $H_0 : \mu = \mu_0$. The following alternative hypotheses can be specified:

- $H_A : \mu < \mu_0$. This is a left-tailed test.
- $H_A : \mu > \mu_0$. This is a right-tailed test.
- $H_A : \mu \neq \mu_0$. This is a two-tailed test.

If a p-value falls within the critical region for the left- and right-tailed tests (to the left or right area of the distribution, respectively), the null hypothesis is rejected. For the same level of significance, say $\alpha = 0.05$, we will "split the difference" for a two-tailed test. If a p-value falls inside the left or right tails in a two-tailed test, the null hypothesis is rejected. Notice the smaller critical regions on both sides of the distribution for a two-tailed test.

4.2.2 The Student's t-distribution

The Student's t-distribution is a continuous probability distribution that assumes a normally distributed population and an unknown standard deviation.

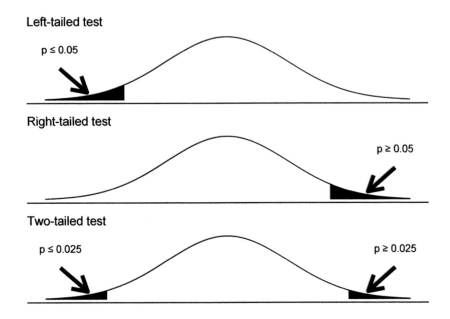

FIGURE 4.2 Conclusions of different alternative hypotheses with a level of significance of 0.05.

The distribution and associated tests were developed by William Sealy Gosset in the early 1900s, a story with a rich history that is told in the popular book *The Lady Tasting Tea* (Salsburg 2001).

The t-distribution is similar in shape to a normal distribution but is less peaked with wider tails. The t-distribution works well with small samples because the sample standard deviation s can be used. (We do not know the population standard deviation σ which differentiates it from the z-distribution.)

The shape of the t-distribution depends on its degrees of freedom. The **degrees of freedom (df)** are the minimum number of independent values needed to specify a distribution. Degrees of freedom vary depending on the number of samples. When we make inference about a population mean μ using a t-distribution, the appropriate degrees of freedom are found by subtracting 1 from the sample size n. Hence $df = n{-}1$. As the degrees of freedom increase, the t density curve more closely resembles the standard normal distribution.

4.2.3 Confidence intervals

Related to the two-tailed hypothesis test and the t-distribution is the confidence interval. A **confidence interval** has a lower and upper bound that

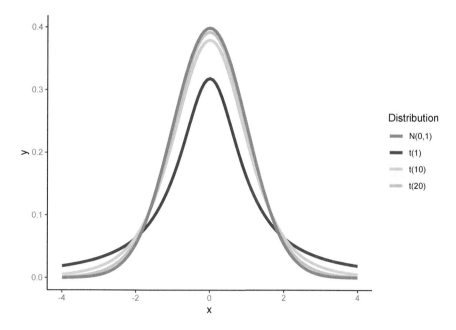

FIGURE 4.3 The Student's t-distribution with different degrees of freedom compared to the standard normal N(0,1) distribution.

contains the true value of a parameter. A 95% confidence level is most commonly used across disciplines, corresponding to a level of significance of $\alpha = 0.05$.

If we sample from a population with an approximate normal distribution, a confidence interval can be calculated as:

$$\bar{y} \pm t_{n-1,\alpha/2} \frac{s}{\sqrt{n}}$$

where $\bar{y}$ is the mean of the sample, $t_{n-1,\alpha/2}$ is the value t from the t-distribution, s is the standard deviation, and n is the number of observations.

In R, the qt() function allows us to calculate the correct value of t to use. This function provides the quantiles of the t-distribution if we specify the quantile we're interested in and the number of degrees of freedom. As an example, we could use the following code to find the t-statistic for the 95th quantile of the t-distribution with 9 degrees of freedom.

```
qt(0.95, df = 9)
```

```
## [1] 1.833113
```

Consider we want to calculate a 95% confidence interval on the number of chirps that a striped ground cricket makes, a data set we saw in Chapter 2 (**chirps**). The data contain the number of chirps per second (cps) and the air temperature measured in degrees Fahrenheit (temp_F):

```
library(stats4nr)

head(chirps)
```

```
## # A tibble: 6 x 2
##     cps temp_F
##   <dbl>  <dbl>
## 1  20     93.3
## 2  16     71.6
## 3  19.8   93.3
## 4  18.4   84.3
## 5  17.1   80.6
## 6  15.5   75.2
```

To find a 95% confidence interval for the number of chirps we will calculate the mean, standard deviation, and number of observations. Notice how we specify the values in the qt() function to provide the appropriate values to obtain the confidence levels:

```
mean_cps <- mean(chirps$cps)
sd_cps<- sd(chirps$cps)
n_cps<- length(chirps$cps)

low_bound <- mean_cps + (qt(0.025,
                           df = n_cps - 1) *
                         (sd_cps / sqrt(n_cps)))
high_bound <- mean_cps + (qt(0.975,
                            df = n_cps - 1) *
                          (sd_cps / sqrt(n_cps)))
low_bound; high_bound
```

```
## [1] 15.71077
```

```
## [1] 17.59589
```

So, we can state that the 95% confidence interval for the number of chirps is [15.71, 17.60]. In application, this means that if we collected another 100 samples of the number of chirps from striped ground crickets, 95 of the means would fall within this confidence interval.

Note that we can change the quantile to calculate a narrower confidence interval. For example, we can use the chirps data to compute a 90% confidence interval by changing the quantile in the qt() function:

```
low_bound_90 <- mean_cps + (qt(0.05,
                            df = n_cps - 1) * (sd_cps / sqrt(n_cps)))
high_bound_90 <- mean_cps + (qt(0.95,
                            df = n_cps - 1) * (sd_cps / sqrt(n_cps)))
low_bound_90; high_bound_90
```

```
## [1] 15.8793
```

```
## [1] 17.42737
```

So, we can state that the 90% confidence interval for the number of chirps is [15.87, 17.43].

DATA ANALYSIS TIP Andrew Gelman, professor of statistics and political science at Columbia University, has reported interesting results about confidence intervals. He has observed that when people are asked to provide levels of uncertainty, their 50% intervals are typically correct 30% of the time. When people provide 90% levels of uncertainty, their intervals are typically correct 50% of the time. This is, in part, why some prefer to use narrower intervals, such as at the 50% level (Gelman 2016).

FIGURE 4.4 A common loon. Image: John Picken.

4.2.4   Exercises

4.1 Use R to calculate the *t*-statistic at the following quantiles with various degrees of freedom.

    a. The 10th quantile with 8 degrees of freedom.
    b. The 10th quantile with 16 degrees of freedom.
    c. The 75th quantile with 10 degrees of freedom.
    d. The 90th quantile with 22 degrees of freedom.

4.2 Mercury is a highly mobile contaminant that can cycle through air, land, and water. The mercury content of 60 eggs were measured from nests of common loons (*Gavia immer*), a diving bird, in New Hampshire, USA (Loon Preservation Committee 1999). They calculated a mean of $\bar{y} = 0.540$ parts per million (ppm) and a standard deviation $s = 0.399$ ppm. Use these summary statistics to answer the following questions.

    a. Use R code to calculate a 95% confidence interval around the mean mercury content of the loon eggs.
    b. Calculate an 80% confidence interval around the mean mercury content of the loon eggs.
    c. Wildlife biologists know that elevated mercury content of loon eggs occurs when mercury content is greater than 0.5 ppm. Reproductive impairment of loon eggs occurs when mercury content is greater than 1.0 ppm. Where do these values fall within both of the confidence intervals you calculated?

4.3 Fisheries biologists measured the weight of yellow perch (*Perca flavescens*), a species of fish. Data were from a sample from a population of yellow perch in a lake. They calculated a mean perch weight of 260 grams and a variance of 10,200 after measuring a sample of 18 perch. Use R code to calculate a 90% confidence interval around the mean perch weight.

4.3   Hypothesis tests for means

4.3.1   One-sample t-test for a mean

A t-statistic can be calculated for a hypothesis test about a population mean μ with unknown population standard deviation σ. The value for the t-statistic represents how many standard errors a sample mean $\bar{y}$ is from a hypothesized mean. In a one-sample t-test for a mean, the t-statistic is calculated by comparing the sample mean $\bar{y}$ to a hypothesized mean μ_0 (in the numerator) and dividing it by the standard error (in the denominator):

$$t = \frac{\bar{y} - \mu_0}{s/\sqrt{n}}$$

In R, the t.test() function performs a number of hypothesis tests related to the t-distribution. For example, assume an entomologist makes a claim that a striped ground cricket makes 18 chirps per second. We can perform a two-sided one-sample t-test at a level of significance of $\alpha = 0.05$ that compares our sample with this value under the following hypotheses:

- The null hypothesis (H_0) states that the true mean of the number of chirps a striped ground cricket makes is **equal to 18**.
- The alternative hypothesis (H_A) states that the true mean of the number of chirps a striped ground cricket makes is **not equal to 18**.

This can be determined by specifying the sample data and the value being tested against in the mu = statement:

```
t.test(chirps$cps, mu = 18)
```

```
##
##   One Sample t-test
##
## data:   chirps$cps
## t = -3.0643, df = 14, p-value = 0.008407
## alternative hypothesis: true mean is not equal to 18
```

```
## 95 percent confidence interval:
##   15.71077 17.59589
## sample estimates:
## mean of x
##   16.65333
```

The calculated t-statistic is -3.0643 with 14 degrees of freedom. The p-value of 0.008407 is less than our level of significance of $\alpha = 0.05$. Hence, we can conclude that we have evidence to reject the null hypothesis and conclude that the true mean of the number of chirps a striped ground cricket makes is not equal to 18.

Also provided in the R output is the mean number of chirps (16.6533) and the 95% confidence interval (15.71077, 17.59589). Our sample mean was less than our hypothesized mean in this one-sample t-test.

The conf.level statement allows you to change the confidence level, corresponding to the complement of the level of significance. For example, we could run the same hypothesis test at a level of significance of 0.10:

```
t.test(chirps$cps, mu = 18, conf.level = 0.90)
```

```
##
##   One Sample t-test
##
## data:  chirps$cps
## t = -3.0643, df = 14, p-value = 0.008407
## alternative hypothesis: true mean is not equal to 18
## 90 percent confidence interval:
##   15.87930 17.42737
## sample estimates:
## mean of x
##   16.65333
```

Notice that while the conclusion of the hypothesis test does not change, the confidence interval is narrower at a more liberal level of significance.

By default the t.test() function will run a two-tailed hypothesis test. To specify a one-sided test, you can use the alternative statement:

- Use alternative = "less" for a left-tailed test when $H_A : \mu < \mu_0$.
- Use alternative = "greater" for a right-tailed test when $H_A : \mu > \mu_0$.

```
t.test(chirps$cps, mu = 18, alternative = "greater")
```

```
##
##   One Sample t-test
##
## data:   chirps$cps
## t = -3.0643, df = 14, p-value = 0.9958
## alternative hypothesis: true mean is greater than 18
## 95 percent confidence interval:
##   15.8793      Inf
## sample estimates:
## mean of x
##   16.65333
```

```
t.test(chirps$cps, mu = 18, alternative = "less")
```

```
##
##   One Sample t-test
##
## data:   chirps$cps
## t = -3.0643, df = 14, p-value = 0.004204
## alternative hypothesis: true mean is less than 18
## 95 percent confidence interval:
##       -Inf 17.42737
## sample estimates:
## mean of x
##   16.65333
```

Notice the differences in outcomes in the two hypothesis tests. The sample mean $\bar{y} = 16.65$ is less that the hypothesized mean $\mu_0 = 18$, so we have evidence to conclude that the true population mean is not greater than 18, but is less than 18. The other important output to note when using the alternative statement is the Inf and -Inf values for the confidence intervals. This is the result of the one-sided test that was specified with no lower and upper bounds.

NONPARAMETRIC STATISTICS: Most of what we discuss in this chapter falls under the family of parametric statistics, which assume that the data sampled can be modeled with a probability distribution. Another family, termed nonparametric statistics or distribution-free methods, make no prior assumption that data reflect a probability distribution. The nonparametric equivalent of the *t*-test is termed the Wilcoxon test. In this test, the distribution is considered symmetric around the hypothesized mean (e.g., 18 with the chirps data). The differences between the data and the hypothesized mean are calculated and ranked to determine the test statistic. With the chirps data, try running `wilcox.test(chirps$cps, mu = 18)` and compare the output with the `t.test()` function. For more on nonparametric statistics, Kloke and McKean (2014) is an excellent reference.

4.3.2   Two-sample t-test for a mean

A two sample *t*-test is used if we want to compare the means of a quantitative variable for two populations. In this case, our parameters of interest are the population means μ_1 and μ_2. Similar to a one-sample test, random samples can be obtained to compare the sample means.

With two-sample hypothesis tests, we'll need to understand the properties of the population standard deviation σ. With two populations, their standard deviations σ_1 and σ_2 may be equal or unequal to one another. That result will influence how the *t*-statistic is calculated.

The primary assumptions before performing a two-sample *t*-test for a mean are as follows:

- The two samples are independent of one another. That is, one observation is not influenced by another.
- The data follow a normal distribution.

There are two versions of *t*-tests that assess the difference between two means: the pooled and Welch *t*-test.

4.3.2.1   Pooled t-test

Assuming the two population standard deviations are equal, a **pooled t-test** calculates a two-sample *t*-statistic as:

$$t = \frac{\bar{y}_1 - \bar{y}_2}{\sqrt{\dfrac{s^2}{n_1} + \dfrac{s^2}{n_2}}}$$

where $\bar{y}_1$, $\bar{y}_2$, n_1, and n_2 represent the means and number of samples for the first and second groups. The variance s^2 is an estimate of the sample variance that is obtained by "pooling" the sample variances from both samples. The degrees of freedom for the test are calculated as $n_1 + n_2 - 2$.

To conduct a two-sample pooled *t*-test, let's add a simulated sample of the number of chirps per second to the **chirps** data set. We do this by using a function from the **tidyverse** function:

```
chirps <- chirps %>%
  add_column(cps_2 = c(20.5, 16.3, 20.9, 18.6, 17.5, 15.7,
                       14.9, 17.5, 15.9, 16.5, 15.3, 17.7,
                       16.2, 17.6, 14.6))
```

After rearranging the data to a long format with `pivot_longer()`, notice the slightly larger values in `cps_2` compared to `cps`:

```
chirps2 <- chirps %>%
  pivot_longer(!temp_F, names_to = "dataset", values_to = "cps")

ggplot(chirps2, aes(dataset, cps)) +
  geom_boxplot() +
  labs(x = "Data set", y = "Chirps per second")
```

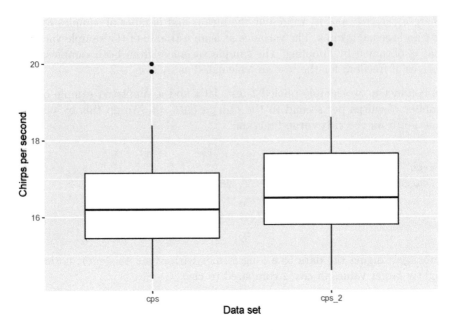

The t.test() function can also be specified for two-sample t-tests. We can specify a two-sided two-sample t-test at a level of significance of $\alpha = 0.05$ that determines whether or not the two populations of ground cricket chirps are equal with the following hypotheses:

- The null hypothesis (H_0) states that the true mean of the number of chirps a striped ground cricket makes in cps is **equal to** cps_2.
- The alternative hypothesis (H_A) states that true mean of the number of chirps a striped ground cricket makes in cps is **not equal to** cps_2.

The following is an example of the pooled t-test assuming variances are equal at a level of significance of $\alpha = 0.05$:

```
t.test(chirps$cps, chirps$cps_2, conf.level = 0.95, var.equal = TRUE)
```

```
##
##   Two Sample t-test
##
## data:  chirps$cps and chirps$cps_2
## t = -0.6034, df = 28, p-value = 0.5511
## alternative hypothesis: true difference in means is not equal to 0
## 95 percent confidence interval:
##  -1.728617  0.941950
```

```
## sample estimates:
## mean of x mean of y
##   16.65333   17.04667
```

Note that the calculated t-statistic is -0.6034 with 28 degrees of freedom. The p-value of 0.5511 is greater than our level of significance of $\alpha = 0.05$. Hence, we have evidence to accept the null hypothesis and conclude that the true mean of the number of chirps a striped ground cricket makes in cps is equal to cps_2.

4.3.2.2 Welch t-test

Assuming the two population standard deviations are not equal, a **Welch t-test** calculates a two-sample t-statistic as:

$$t = \frac{\bar{y}_1 - \bar{y}_2}{\sqrt{\frac{s_1^2}{n_1} + \frac{s_2^2}{n_2}}}$$

where $\bar{y}_1$, $\bar{y}_2$, s_1^2, s_2^2, n_1, and n_2 represent the means, variances, and number of samples for the first and second groups. Notice how the variances from each sample are used directly in this calculation. The degrees of freedom for the Welch test is a more complex calculation using the variance and number of observations from each sample.

We can assume the same hypothesis test comparing means of cps and cps_2. The following is an example of the Welch t-test assuming variances are not equal at a level of significance of $\alpha = 0.05$:

```
t.test(chirps$cps, chirps$cps_2, conf.level = 0.95, var.equal = FALSE)
```

```
##
##   Welch Two Sample t-test
##
## data:  chirps$cps and chirps$cps_2
## t = -0.6034, df = 27.77, p-value = 0.5511
## alternative hypothesis: true difference in means is not equal to 0
## 95 percent confidence interval:
##   -1.7291150  0.9424483
## sample estimates:
## mean of x mean of y
##   16.65333   17.04667
```

Note that the *t*-statistic and *p*-values are the same for both tests. However, the degrees of freedom are slightly lower in the Welch *t*-test (27.77) compared to the pooled *t*-test (28). For this relatively small data set, this results in slight differences in the lower and upper bounds of the confidence intervals. Before deciding whether to use the pooled or Welch test, consult the procedures outlined in the *Hypothesis tests for variances* section later in this chapter.

4.3.3   Paired t-test

Two samples are often dependent on one another, that is, one observation is influenced by the other. A **paired *t*-test** evaluates the difference between the responses of two paired measurements.

When paired data result from measuring the same quantitative variable twice, the paired *t*-test analyzes the differences in each pair. For example, consider a survey designed to examine hunter attitudes to new hunting policies. Ten hunters take a survey on their knowledge of new policies, then listen to a presentation that describes the new policies in detail. After the presentation the hunters take a similar survey on their knowledge. In this example a paired *t*-test is appropriate because the two samples are dependent, that is, the same hunters are being surveyed.

The paired *t*-test is similar in design to a one-sample *t*-test. Instead, the differences of the two means are evaluated. We will first find the mean difference ($\bar{d}$) between the responses within each pair $\bar{y}_1$ and $\bar{y}_2$. Then we will calculate the test statistic to evaluate these differences:

$$t = \frac{\bar{d} - \mu_d}{s_d/\sqrt{n}}$$

where μ_d is a value to be tested against (typically 0), s_d is the standard deviation of the difference, and n is the number of pairs.

For example, assume a group of eight hunters were first asked how many days they would spend afield during a hunting season. Then, the same hunters listened to a presentation by a wildlife biologist describing new hunting regulations that would adjust the number of opening and closing dates of the hunting season by geographic regions, ultimately leading to a longer hunting season. The hunters again estimated how many days they would spend afield after the new regulations took place. Here are the data represented in the days variables:

```
hunt <- tribble(
  ~hunterID, ~time, ~days,
  1, " Pre", 5,
```

```
   2, " Pre", 7,
   3, " Pre", 3,
   4, " Pre", 9,
   5, " Pre", 4,
   6, " Pre", 1,
   7, " Pre", 6,
   8, " Pre", 2,
   1, "Post", 8,
   2, "Post", 7,
   3, "Post", 4,
   4, "Post", 11,
   5, "Post", 9,
   6, "Post", 3,
   7, "Post", 10,
   8, "Post", 5
)

ggplot(hunt, aes(time, days)) +
  geom_boxplot() +
    geom_point() +
  labs(x = "Time period",
       y = "Estimated days to hunt")
```

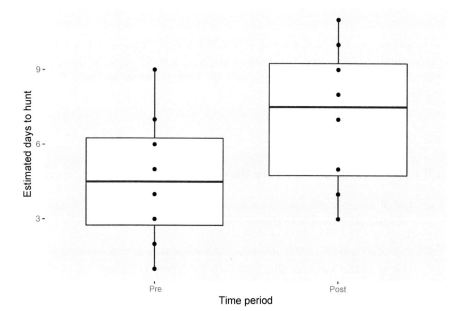

Our null hypothesis can be written as $H_0 : \mu_d = 0$ and alternative hypothesis $H_A : \mu_d \neq 0$. After filtering the pre and post measurements, we can specify the `paired = TRUE` statement to the `t.test()` function to perform a paired *t*-test:

```
pre <- hunt %>%
  filter(time == " Pre")

post <- hunt %>%
  filter(time == "Post")

t.test(pre$days, post$days, paired = TRUE)



##
##   Paired t-test
##
## data:  pre$days and post$days
## t = -4.4096, df = 7, p-value = 0.00312
## alternative hypothesis: true difference in means is not equal to 0
## 95 percent confidence interval:
##   -3.840616 -1.159384
## sample estimates:
## mean of the differences
##                     -2.5
```

The *t*-statistic is -4.41 and *p*-value is 0.00312 with 7 degrees of freedom. On average, hunters estimated they would spend 2.5 more days in the field after listening to the presentation. Hence, we have evidence to reject the null hypothesis.

4.3.4   Exercises

4.4 For the following questions, use the **elm** data set.

   a. Use R to calculate the mean, standard deviation, number of observations, and standard error of the mean for the mean of tree height (HT).
   b. Calculate a 95% confidence interval for the population mean of tree height (HT) from the elm data set. Interpret the meaning of the confidence interval and what it is telling you.

4.5 An urban forester working with cedar elm trees mentions that tree height for this species is around 30 feet.

    a.   Perform a two-sided one-sample t-test at a level of significance of $\alpha = 0.05$ comparing the HT variable with this value. What are the null and alternative hypotheses for this test? How would you interpret the results?

    b.   Perform a one-sided one-sample t-test at a level of significance of $\alpha = 0.10$ with the null hypothesis that the height of the elm trees is less than 30 feet and the alternative hypothesis that the height of the elm trees is equal to 30 feet. How would you interpret the results?

4.6 For this next question, we will perform a series of two-sample t-tests comparing intermediate and suppressed trees in the **elm** data set.

    a.   Use the `filter()` function and the CROWN_CLASS_CD variable to create two new data sets: one containing the intermediate trees and one containing the suppressed trees. How many observations are there for each data set?

    b.   Perform a two-sample t-test at a level of significance of $\alpha = 0.05$ with the null hypothesis that the height for intermediate trees is equal to the height for suppressed trees and the alternative hypothesis that the heights are not equal. Assume the variances of the two height samples are not equal. How would you interpret the results?

    c.   Perform a two-sample t-test at a level of significance of $\alpha = 0.10$ with the null hypothesis that the diameters (DIA) for intermediate trees is equal to the diameters for suppressed trees and the alternative hypothesis that the diameters are not equal. Assume the variances of the two diameter samples are equal. How would you interpret the results?

4.7 With the **hunt** data set, we performed a paired t-test that examined changes in the number of days that hunters would spend afield. Imagine that the Pre and Post estimates were independent. Conduct a two-sample t-test that pools the variances of both samples. What are the primary differences in the hypothesis test performed at a level of significance of $\alpha = 0.05$?

4.4 Hypothesis tests for variances

4.4.1 The *F*-distribution

Just like testing for the mean, we might be interested in knowing if the *variances* of two populations are equal. This could help in assessing whether or not we can use a pooled *t*-test for the means, which assumes that the variances of two samples are equal.

The *F*-distribution is used when conducting a hypothesis test for variances. The *F*-distribution has the following characteristics:

- It is a non-negative, continuous distribution.
- The majority of the area of the distribution is near 1.
- It peaks to the right of zero. The greater the *F* value, the lower the curve.

The *F*-distribution depends on two values for the degrees of freedom (one for each sample), where $df_1 = n_1 - 1$ and $df_2 = n_2 - 1$. The number of samples results in different shapes of the *F*-distribution:

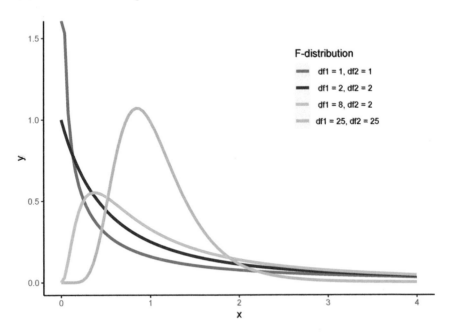

FIGURE 4.5 The F-distribution with varying degrees of freedom collected from two samples.

In practice, consider an F-distribution with 4 and 9 degrees of freedom. The upper-tail value from the $F_{4,9}$ distribution so that the area under the curve to the right of it is 0.10 can be found with the qf() function:

```
qf(p = 0.90, df1 = 4, df2 = 9)
```

```
## [1] 2.69268
```

So, with 4 and 9 degrees of freedom, 10% of the area under the curve is greater than 2.69.

4.4.2 Two-sample F-test for variance

Assuming independent samples, our null hypothesis for the F-test for variances can be written as $H_0 : \sigma_1^2 = \sigma_2^2$. Similar to a two-sample hypothesis for means, the following alternative hypotheses can be specified:

- $H_A : \sigma_1^2 < \sigma_2^2$
- $H_A : \sigma_1^2 > \sigma_2^2$
- $H_A : \sigma_1^2 \neq \sigma_2^2$

We can assume the sample variance s_1^2 estimates the population variance σ_1^2 and similarly s_2^2 estimates σ_2^2. A test statistic F_0 is then calculated using the ratio of these two values, that is:

$$F_0 = \frac{s_1^2}{s_2^2}$$

In R, the var.test() function performs an F-test to compare two variances. As an example, consider the data from the number of chirps that a striped ground cricket makes. These data are from the **chirps** data set used earlier in the chapter. Our goal is to test the hypothesis that the variances for the cps and cps_2 variables are equal at a level of significance of $\alpha = 0.10$:

```
var.test(chirps$cps, chirps$cps_2, conf.level = 0.90)
```

```
##
##  F test to compare two variances
##
## data:  chirps$cps and chirps$cps_2
## F = 0.83319, num df = 14, denom df = 14, p-value = 0.7375
```

```
## alternative hypothesis: true ratio of variances is not equal to 1
## 90 percent confidence interval:
##   0.3354586 2.0694086
## sample estimates:
## ratio of variances
##             0.8331872
```

With a resulting p-value of 0.7375, we accept the null hypothesis and conclude that the variances for cps and cps_2 are equal. The ratio between variances of 0.8331 is close to 1, indicating the variances for both samples were similar. This result gives us confidence that we can use a pooled two-sample t-test that assumes equal variances between the two samples. It is a good practice to first conduct an F-test for variances and then conduct a two-sample t-test that either assumes the variances are equal (the pooled test) or not equal (the Welch test).

4.4.3   Exercises

4.8 Consider an F-distribution with the following upper-tail values and degrees of freedom. Use the qf() function to find the values of F that correspond to the parameters. What do you notice about the F values as the degrees of freedom and areas change?

    a.   The area under the curve to the right of the value is 0.20 with 4 and 6 degrees of freedom.

    b.   The area under the curve to the right of the value is 0.05 with 4 and 6 degrees of freedom.

    c.   The area under the curve to the right of the value is 0.10 with 5 and 10 degrees of freedom.

    d.   The area under the curve to the right of the value is 0.10 with 10 and 20 degrees of freedom.

4.9 For this next question, consider the intermediate and suppressed trees from the **elm** data set and your analyses from question 4.6.

    a.   Perform an F-test comparing the variances for the heights (HT) of intermediate and suppressed trees using a level of significance of $\alpha = 0.05$. How would you interpret the results?

    b.   Perform an F-test comparing the variances for the diameters (DIA) of intermediate and suppressed trees using a level of significance of $\alpha = 0.10$ How would you interpret the results?

    c.   You will find a similar outcome for the hypothesis tests in the previous two questions. If you were interested in following up your analysis with a two-sample t-test of means, is a pooled or Welch t-test appropriate?

4.5   Summary

Statistical tests enable us to compare observed data with a claim whose truth we want to assess. We may be interested in seeing if a mean differs from a unique value (a one-sample t-test), if two independent means differ (a two-sample t-test), or if two dependent means differ (a paired t-test). We can also compare whether or not the variances from two samples differ. This can lead to a decision about which test for means is appropriate, i.e., a pooled or Welch t-test.

In these statistical tests, alternative hypotheses can be set up to be not equal to, less than, or greater than a specified value. The choice on how to set up the hypothesis test depends on the question being asked of the data. Two distributions were introduced in this chapter: the t-distribution and F-distribution. These distributions are widely used in other topics in statistics such as regression and analysis of variance, so it's important to understand them as we'll revisit them in future chapters.

4.6   References

Gelman, A. 2016. Why I prefer 50% rather than 95% intervals. Available at: https://statmodeling.stat.columbia.edu/2016/11/05/why-i-prefer-50-to-95-intervals/

Kloke, J., J.W. McKean. *Nonparametric statistical methods using R.* Chapman and Hall/CRC Press. 287 p.

Loon Preservation Committee. 1999. Mercury exposure as measured through abandoned common loon eggs in New Hampshire, 1998. Available at: https://www.fws.gov/newengland/PDF/contaminants/Contaminant-Studies-Files/Mercury-Exposure-as-Measured-through-Abandoned-Common-Loon-Eggs-in-New-Hampshire-1998-1999.pdf

Salsburg, D. 2001. *The lady tasting tea: how statistics revolutionized science in the twentieth century.* Holt Publishing. 340 pp.

Wasserstein, R.L., and N.A. Lazar. 2016. The ASA statement on p-values: context, process, and purpose. *The American Statistician* 70(2): 129–133.

5

Inference for counts and proportions

5.1 Introduction

One of the most common forms of quantitative data in natural resources are counts and proportions. This might include whether or not a species is present or absent in an area, whether a plant survived or died after an experimental treatment, or whether a rain event led to flood conditions or not.

In Chapter 3 we learned about the Bernoulli and binomial distributions, two discrete distributions that are common across natural resources. Counts of events and distributions can be modeled with the binomial distribution, so we will continue to use them in hypothesis tests with data collected as counts and proportions. What will make our jobs easier is to apply a normal approximation to make inferences about the proportion of successes in a population. When working with counts and proportions, it is critical to use the correct methods and analyses in R to make the appropriate inference on the population of interest.

5.2 The normal approximation for a binomial distribution

When the same chance process is repeated several times, we are often interested in whether a particular outcome happens on each repetition. In some cases, the number of repeated trials is fixed in advance and we are interested in the number of times a particular event (or a "success") occurs. Recall from Chapter 3 that a binomial distribution is a series of Bernoulli trials n with probability of success p. Hence, the mean and variance of the binomial distribution is np and $np(1\text{-}p)$, respectively.

Listed below are the four conditions of a binomial distribution. Moore et al. (2017) propose to think of **BINS** as a reminder of these conditions:

- **Binary.** All possible outcomes of each trial can be classified as "success" or "failure."
- **Independent.** Trials must be independent; that is, knowing the result of one trial must not have any effect on the result of another.
- **Number.** The number of trials n of the chance process must be fixed in advance.
- **Success.** In each trial, the probability p of success must be the same.

The normal distribution allows us to make inference using several procedures and assumptions outlined in previous chapters. Fortunately, a normal approximation can be applied to a binomial distribution if n is relatively large. If $np \geq 10$ and $n(1-p) \geq 10$, we can apply the normal approximation to the binomial distribution.

As an example, consider that you are tasked with inspecting company facilities that store hazardous waste as a part of their operations. The probability of issuing a violation to the company for not following hazardous waste storage regulations is $p = 0.24$. We can confirm the data are binomial in their design: each facility is issued a violation or not (binary), inspections do not rely on the outcome of other inspections (independent), the number of inspections is fixed (as seen in the example below), and the probability of "success" is the same (e.g., the 24% chance a company is issued a violation).

Imagine you inspect 8, 20, and 80 facilities that result in the following distributions:

Note that when n increases from 8 to 80, the distribution begins to approximate a normal distribution. We can determine the normal distribution approximation can be used in each of the scenarios:

- For $n = 8$, $np = 8 * 0.24 = 1.9$ and $n(1-p) = 8 * (1 - 0.24) = 6.0$.
- For $n = 20$, $np = 20 * 0.24 = 4.8$ and $n(1-p) = 20 * (1 - 0.24) = 15.2$.
- For $n = 80$, $np = 80 * 0.24 = 19.2$ and $n(1-p) = 80 * (1 - 0.24) = 60.8$.

Hence, only for $n = 80$ can we apply the normal approximation because $np \geq 10$ and $n(1-p) \geq 10$.

5.2.1 Exercises

5.1 Write a function in R that determines whether or not you can apply the normal distribution approximation for a given set of parameters n and p from a binomial distribution. Then, use that function to determine whether or not you can apply the approximation based on the following scenarios.

 a. You plant 10 tree seedlings with a known survival rate 85% after the first growing season.

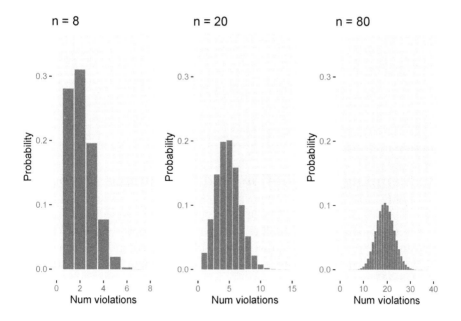

FIGURE 5.1 Hazardous waste violation probabilities with n = 8, 20, and 80.

b. You plant 90 tree seedlings with a known survival rate 85% after the first growing season.
c. You sample 20 ruffed grouse with a probability having West Nile virus of 0.13.
d. There is a 40% chance of hearing a golden-winged warbler on a birding trip. You make 25 visits.

5.2 The code below uses ggplot() to produce a plot of the number of hazardous waste violations when $n = 80$ and $p = 0.24$. Change the arguments in the code to produce two plots showing the distribution of survival of 10 and 90 tree seedlings, each with a known survival rate of 85% after the first growing season. In particular, modify the scale_x_continuous() and scale_y_continuous() arguments to "zoom in" on the distribution.

```
x.80 <- 0:80
df.80 <- tibble(x = x.80, y = dbinom(x.80, 80, 0.24))

p.80 <- ggplot(df.80, aes(x = x.80, y = y)) +
  geom_bar(stat = "identity") +
  scale_x_continuous(limits = c(0, 40)) +
```

```
scale_y_continuous(limits = c(0, 0.35)) +
labs(title = "n = 80",
     x = "Number of violations",
     y = "Probability")
```

5.3 Sampling distribution and one-sample hypothesis test

Just like we looked at the sampling distribution of a mean value in Chapter 4, we can look at the sampling distribution of a proportion. Here, the sample statistic $\hat{p}$ is used to estimate p, the population proportion. The standard deviation of $\hat{p}$ is

$$\sigma_{\hat{p}} = \sqrt{\frac{p(1-p)}{n}}$$

Since we're interested in the population parameter p, we can replace p with $\hat{p}$ to find the standard error of the sample proportion:

$$\sigma_{\hat{p}} = \sqrt{\frac{\hat{p}(1-\hat{p})}{n}}$$

For sample proportions, we use the z-distribution for making inference about population parameters:

The null hypothesis for a one=sample proportion test can be written as $H_0 : p = p_0$, where p_0 is a proportion we're testing against. The following alternative hypotheses can be specified in a one-sample test for a proportion:

- $H_A : p < p_0$ (left-tailed test)
- $H_A : p > p_0$ (right-tailed test)
- $H_A : p \neq p_0$ (two-tailed test)

Using the z-distribution, we calculate a statistic for this hypothesis test as:

$$z = \frac{\hat{p} - p}{\sqrt{\frac{p_0(1-p_0)}{n}}}$$

In R, two functions perform one-sample hypothesis tests for proportions. The

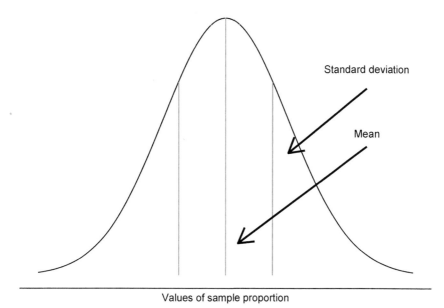

Standard deviation

Mean

Values of sample proportion

FIGURE 5.2 The sampling distribution of a sample proportion.

`prop.test()` function applies the normal approximation to the binomial distribution and should be used when those conditions are met. The `binom.test()` is designed for smaller samples and should be used when the normal approximation conditions are not met.

For example, consider an organization that wants to make the claim that the majority of farmers (>50% of them) support limiting their use of nitrogen in their practices. While adding nitrogen in the form of fertilizer can improve the growth of crops, it can also run off into unintended water bodies. A polling company surveyed 30 farmers and their results indicated 18 of them supported limiting their use of nitrogen.

We will conduct a hypothesis test with the null hypothesis $H_0 : p = 0.5$ and the alternative hypothesis $H_A : p > 0.5$. The test will use a level of significance $\alpha = 0.10$.

Our proportion of successes is $p = 18/30 = 0.6$, so because $np = 30 * 0.6 = 18$ and $n(1 - p) = 30 * (1 - 0.6) = 12$, the normal distribution approximation is confirmed and the `prop.test()` function can be used. To carry out this one-sided hypothesis test, we enter x = 18 successes (number of farmers that agree with the limitations), n = 30 total farmers, a null proportion p = 0.50, and alternative = "greater" because we're examining if the majority of farmers agree with the limitations:

FIGURE 5.3 Application of nitrogen fertilization on a row crop. Image: University of Minnesota Extension.

```
prop.test(x = 18, n = 30, p = 0.50,
          alternative = "greater", conf.level = 0.90)
```

```
##
##   1-sample proportions test with continuity correction
##
## data:  18 out of 30, null probability 0.5
## X-squared = 0.83333, df = 1, p-value = 0.1807
## alternative hypothesis: true p is greater than 0.5
## 90 percent confidence interval:
##  0.4666054 1.0000000
## sample estimates:
##   p
## 0.6
```

Results indicate a *p*-value of 0.1807, indicating the survey has little evidence to support the statement that the majority of farmers support limits of nitrogen use. Thus, we accept the null hypothesis. The output also provides a chi-squared value of 0.83333. For comparison, we can also input the same statistics using the `binom.test()` function:

```
binom.test(x = 18, n = 30, p = 0.50,
           alternative = "greater", conf.level = 0.90)
```

```
##
##  Exact binomial test
##
## data:  18 and 30
## number of successes = 18, number of trials = 30, p-value = 0.1808
## alternative hypothesis: true probability of success is greater than 0.5
## 90 percent confidence interval:
##  0.4665663 1.0000000
## sample estimates:
## probability of success
##                    0.6
```

Note the *p*-values and lower bound for the confidence interval are slightly different in both functions.

5.3.1 Exercises

5.3 Perform the following hypothesis tests for proportions at a level of significance $\alpha = 0.05$ using the `prop.test()` function and the given parameters. State your conclusions in a sentence or two.

    a. In a survey of 600 residents, 54% indicated they supported mining in their county. Perform a test that examines if your results can state that the majority of residents support mining.

    b. A tree nursery ensures that planting their seedlings will result in at least 90% survival after two years of growth. You question these claims after measuring 250 seedlings and find 215 of them survived after two years. Perform a test that examines if your results support the nursery's claims.

    c. You sample 54 ruffed grouse for West Nile virus and find 5 of them to have the disease. Historical data indicate you expect 13% of grouse to test positive for the virus. You want to make the claim that you observed a positive infection rate less than the historical average.

5.4 The Random Geographic Coordinates web page (`https://www.random.o rg/geographic-coordinates/`) uses randomness to select a set of geographic coordinates at any location on Earth. Use the website to generate a random coordinate and record whether it falls on *land* or *water*. Repeat this so that you have 20 random locations and whether they are on land or water.

Determine the proportion of Earth covered with water from your collection of 20 random samples. Now, do an internet search to find the true proportion of Earth covered in water. Use `prop.test()` to run two two-tailed hypothesis tests for proportions, one at 90% and the other at a 50% confidence level. The null hypothesis is that your sampled proportion is equal to the true proportion of Earth covered in water. Does the true proportion of Earth covered in water fall within the calculated confidence interval provided in the R output?

5.4 Two-sample hypothesis tests for proportions

Remember that when we switched from using the z to the t-distribution, we began using the sample standard deviation s as opposed to the population standard deviation σ. Similar to two-sample hypothesis tests for means, proportions use the properties of the population standard deviation σ as the basis for their inference.

Two-sample hypothesis tests for proportions examine p_1 and p_2, two population parameters representing different groups. The difference between the two proportions is the value that is examined in these hypothesis tests. When data are sampled randomly from two groups, we can approximate the difference $p_1 - p_2$ with $\hat{p}_1 - \hat{p}_2$, each of which may have different sample sizes n_1 and n_2.

The z-distribution can similarly be used for making inference about the difference in sample statistics $\hat{p}_1 - \hat{p}_2$. The sample standard deviation of can be estimated by:

$$\sigma_{\hat{p}_1 - \hat{p}_2} = \sqrt{\frac{p_1(1-p_1)}{n_1} + \frac{p_2(1-p_2)}{n_2}}$$

We need a way to estimate a combined or "pooled" proportion p, so we can combine the data from the two samples by dividing the number of successes in both samples by the total number collected by both samples and call it $\hat{p}$.

The null hypothesis for a two-sample test for proportions can be written as $H_0 : p_1 = p_2$. The z-statistic for this hypothesis test is:

$$z = \frac{\hat{p}_1 - \hat{p}_2}{\sqrt{\hat{p}(1-\hat{p})(\frac{1}{n_1} + \frac{1}{n_2})}}$$

The following alternative hypotheses can be specified in two-sample tests for proportions:

- $H_A : p_1 - p_2 < 0$ (left-tailed test)
- $H_A : p_1 - p_2 > 0$ (right-tailed test)
- $H_A : p_1 - p_2 \neq 0$ (two-tailed test)

The `prop.test()` function can also perform hypothesis tests that compare two sample proportions.

For example, consider an organization that wants to compare farmer support in limiting nitrogen use in two counties. In one county, a polling company surveyed 40 farmers and 19 of them supported limiting their use of nitrogen. In a different county, a polling company surveyed 36 farmers and 24 of them supported limited nitrogen use. We can specify these data as vectors:

```
farmer.support <- c(19, 24)
farmer.total <- c(40, 36)
```

So, our proportion of farmers that support limiting their nitrogen use is $p_1 = 19/40 = 0.48$ and $p_2 = 24/36 = 0.66$ in the first and second counties, respectively. We will conduct a hypothesis test with the null hypothesis $H_0 : p_1 = p_2$ and the alternative hypothesis $H_A : p_1 - p_2 \neq p_0$. The test will use a level of significance $\alpha = 0.10$.

To carry out this two-sided hypothesis test, we enter both of the vectors that contain the data as the first two arguments:

```
prop.test(farmer.support, farmer.total,
          alternative = "two.sided", conf.level = 0.90)
```

```
##
##  2-sample test for equality of proportions with continuity correction
##
## data:  farmer.support out of farmer.total
## X-squared = 2.1068, df = 1, p-value = 0.1466
## alternative hypothesis: two.sided
## 90 percent confidence interval:
##  -0.40127198  0.01793865
## sample estimates:
##    prop 1    prop 2
## 0.4750000 0.6666667
```

Results indicate a *p*-value of 0.1466, indicating no difference between counties in farmer support for limiting nitrogen use. So, we accept the null hypothesis. Note that the 90% confidence interval provides upper and lower bounds around the mean difference between both proportions (0.1917).

When two samples contain either a success or failure, we can represent this in the form of a 2x2 table. The `matrix()` function allows us to view the data

Inference for counts and proportions

in a compact format. The column and row name labels can be renamed with
`colnames()` and `rownames()`, respectively:

```
farmer.nitrogen <- matrix(c(19, 24, 21, 12), ncol = 2, byrow = T)
colnames(farmer.nitrogen) <- c("Support", "DoNotSupport")
rownames(farmer.nitrogen) <- c("County1", "County2")
farmer.nitrogen
```

```
##         Support DoNotSupport
## County1      19           24
## County2      21           12
```

For data in a 2x2 table like this, the `chisq.test()` function will provide the
same results that `prop.test()` does. Note the minimal output provided using `chisq.test()` compared to the `prop.test()` function (e.g., no confidence
interval or means of sample estimates):

```
chisq.test(farmer.nitrogen)
```

```
##
##   Pearson's Chi-squared test with Yates' continuity correction
##
## data:  farmer.nitrogen
## X-squared = 2.1068, df = 1, p-value = 0.1466
```

For small samples and low numbers of successes (typically fewer than five), the
`fisher.test()` function is appropriate for comparing two sample proportions.
This function also requires the data to be in a matrix format:

```
fisher.test(farmer.nitrogen,
            alternative = "two.sided", conf.level = 0.90)
```

```
##
##   Fisher's Exact Test for Count Data
##
## data:  farmer.nitrogen
## p-value = 0.1092
## alternative hypothesis: true odds ratio is not equal to 1
## 90 percent confidence interval:
```

```
##   0.187029 1.091269
## sample estimates:
## odds ratio
##   0.4572285
```

Note the difference in the *p*-value for both tests (0.1466 and 0.1092) and the way in which the confidence interval differs. For the `fisher.test()` function, a confidence interval for the **odds ratio** is provided that compares the likelihood of p_1 with p_2. The 90% confidence interval for the odds ratio contains 1, indicating the proportions between farmer support of nitrogen use in each county is similar.

5.4.1 Exercises

5.5 Perform the following two-sample hypothesis tests for proportions at a level of significance $\alpha = 0.05$ using the `prop.test()` or `chisq.test()` functions and the given parameters. State your conclusions in a sentence or two.

a. In a survey of 500 residents in each of two counties, 280 and 220 residents in each county indicated they supported mining. Perform a test that examines whether or not the proportion of support for mining differs in these counties.

b. After planting 250 oak tree seedlings you find 215 of them survived after two years. After planting 175 maple tree seedlings you find 165 of them survived after two years. Perform a test that examines whether or not the proportion of surviving oak trees is less than the proportion of surviving maple trees.

c. You sample 64 ruffed grouse for West Nile virus in Michigan and find 8 of them to have the disease. In a similar study in Wisconsin, 45 grouse were tested and 7 had the disease. Perform a test that examines whether or not the proportion of West Nile infections differs in these two states.

5.6 The air quality data set is a built-in data set in R that provides daily air quality measurements collected in New York City from May to September 1973.

Run the code below to read in the air quality data set. We'll name the data set **air** using the `tibble()` function:

```
library(tidyverse)
air <- tibble(airquality)
```

Two variables in `air` that we will use are:

- Mean ozone in parts per billion from 1300 to 1500 hours at Roosevelt Island (`Ozone`)
- Numeric month from May through September (`Month`)

Write R code to find the number of days when the ozone level was unhealthy for sensitive groups or greater, defined as an ozone measurement greater than or equal to 101. Then, perform a two-sample test of proportions that determines if the proportion of unhealthy air quality days was different between the spring (months of May and June) and summer (months of July, August, and September). Perform the tests at a level of significance $\alpha = 0.05$. NOTE: Do not include days with an `NA` for ozone measurement.

5.5   Summary

The general approach to hypothesis tests using proportions is similar to how we handled continuous data. The primary difference in the test for proportions is the use of the z rather than the t-statistic. Applying the normal approximation for a binomial distribution will allow us to make inference with data stored as counts and proportions. The `prop.test()` function allows us to perform many one- and two-sample hypothesis tests for proportions in R.

We'll return to these concepts in Chapter 6 when we discuss chi-squared tests more in depth. There, you might imagine that we have more rows and columns, for example, if we asked farmers from three counties their opinion about nitrogen use and they could respond by stating that they support, do not support, or have no opinion on the topic (resulting in a 3x3 table). In the future, we will calculate the appropriate sample size that allows us to estimate a population proportion within a given margin of error, a calculation that has long been used in political and opinion polling.

5.6   Reference

Moore, D.S., G.P. McCabe, B.A. Craig. 2017. *Introduction to the practice of statistics, 9th ed.* W.H. Freeman and Company, New York.

6

Inference for two-way tables

6.1 Introduction

If data are collected as counts with two outcomes and two groups (e.g., the presence/absence of a species in two different regions), we can perform hypothesis tests for proportions. What adds more complexity is when it is possible to have more than two outcomes or groups being tested against, e.g., the presence/absence of a species in *four* different regions. Count data are quantitative, and a categorical data analysis technique will allow us to properly analyze these data across groups.

Storing data in two-way tables will allow us to visualize these counts across groups in an efficient way. Two-way tables and their analysis are common when experimenting with survey data and categorical data in natural resources. The chi-squared test and the statistic it provides will measure how far the observed counts are from what we expect them to be, ultimately telling us about the association across groups. As a result of this, our hypothesis test will reflect whether or not we can say that the groups are independent from one another.

6.2 Data in two-way tables

Two-way tables can show a lot of information in a compact form. A lot of published data such as survey results are summarized in this format. As an example, we might ask citizens whether or not they approve of the performance of a political candidate (yes or no) by their political party (e.g., liberal or conservative).

A few characteristics of two-way tables are that they:

- describe two categorical variables,
- organize counts in rows and columns, and

- show the number of counts within each cell.

The rows of a two-way table are the values of one categorical variable, and the columns are the values of the other categorical variable. The count in any particular cell of the table equals the number of subjects who fall into that cell.

Consider an example with diseased ruffed grouse. In 2018, a multi-state effort examined the presence of West Nile virus in ruffed grouse populations across the US Lake States. The study concluded that while the West Nile virus was present in the region, grouse that were exposed to the virus could survive.

The total number of ruffed grouse sampled that showed antibodies consistent with West Nile virus is shown by creating the two-way table below. The `matrix()` function creates a table with three states and the number of grouse that tested positive or negative for the virus:

```
grouse <- matrix(c(185, 28, 239, 34, 167, 68), nrow = 3, byrow = T)
colnames(grouse) <- c("West Nile Negative", "West Nile Positive")
rownames(grouse) <- c("Michigan", "Minnesota", "Wisconsin")
grouse
```

```
##           West Nile Negative West Nile Positive
## Michigan                 185                 28
## Minnesota                239                 34
## Wisconsin                167                 68
```

We can see that Minnesota had the greatest number of negative West Nile cases (239) and Wisconsin had the greatest number of positive West Nile cases (68). Visualizing data in this two-way table will help us to set up and interpret the results of a chi-squared test.

It also might help to add a total column to grouse that calculates the total number of grouse for each state and negative/positive outcome. This can be accomplished with the `addmargins()` function:

```
addmargins(grouse)
```

```
##           West Nile Negative West Nile Positive Sum
## Michigan                 185                 28 213
## Minnesota                239                 34 273
## Wisconsin                167                 68 235
## Sum                      591                130 721
```

We can easily see that 721 total grouse were sampled across the three states and 130 of them tested positive for West Nile.

6.3   The chi-squared test for two-way tables

The goal of a **chi-squared test** is to test the null hypothesis (H_0) that there is no relationship between the two categorical variables. In the grouse example, H_0 would be that there is no relationship between positive West Nile cases and the state where they were sampled.

To perform a chi-squared test, we compare the *actual* counts (from the sample data) with *expected* counts (those expected when there is no relationship between the two variables). The expected count in any cell of a two-way table when H_0 is true is:

$$\text{expected count} = \frac{\text{row total} \cdot \text{column total}}{n}$$

With r rows and c columns in a two-way table, there are $r \cdot c$ possible cells. The **chi-squared statistic**, denoted χ^2, measures how far the actual counts are from the expected counts. In this case, "actual" counts are from the sample data and "expected" counts represent the expected count for the same cell. After summing all values in each $r \cdot c$ cell in the table, the result is the chi-squared statistic:

$$\chi^2 = \Sigma \frac{(\text{Actual - Expected})^2}{\text{Expected}}$$

When the actual counts are *very different* from the expected counts, a large value of χ^2 will result. This would provide evidence against the null hypothesis (i.e., we would reject the null). When the actual and expected counts are in *close agreement*, a small value of χ^2 will result. This would provide evidence in favor of the null hypothesis (i.e., we would fail to reject the null).

The primary assumptions of using the χ^2 distribution are that data are obtained from a simple random sample and that each observation can fall into only one cell in the two-way table. The χ^2 distribution has the following properties:

- it contains only positive values,
- its distribution is right-skewed, and
- its degrees of freedom are $(r-1)(c-1)$.

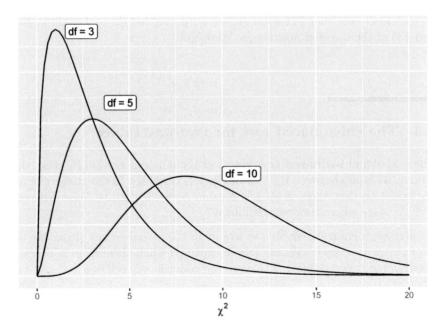

FIGURE 6.1 The chi-squared distribution with different degrees of freedom.

Changing the degrees of freedom leads to many different shapes of the χ^2 distribution. Increasing the degrees of freedom results in a distribution that is less skewed and more bell-shaped:

The calculated proportions within a two-way table represent the conditional distributions describing the relationships between both variables. In a chi-squared test, you describe the relationship by comparing the conditional distributions of the response variable (e.g., West Nile infection) for each level of the explanatory variable (e.g., the state where grouse were sampled). The hypotheses for the grouse example are then:

H_0: West Nile infection and the state where grouse were sampled are independent.

H_A: There is a relationship between West Nile infection and the state where grouse were sampled.

Before conducting the χ^2 test, we may be interested in calculating descriptive statistics that convey important information in the table. As an example, these can be the column or row percentages. A different number of grouse were sampled in each state, making it difficult to see trends across the three states. The `prop.table()` function turns counts into proportions:

```
prop.table(grouse)
```

```
##              West Nile Negative West Nile Positive
## Michigan             0.2565881         0.03883495
## Minnesota            0.3314840         0.04715673
## Wisconsin            0.2316227         0.09431345
```

The `addmargins()` function will sum the proportions within each row and column. This makes it easier to see that nearly 82% of grouse test negative for West Nile and Minnesota had the largest number of grouse tested:

```
addmargins(prop.table(grouse))
```

```
##              West Nile Negative West Nile Positive       Sum
## Michigan             0.2565881         0.03883495 0.2954230
## Minnesota            0.3314840         0.04715673 0.3786408
## Wisconsin            0.2316227         0.09431345 0.3259362
## Sum                  0.8196949         0.18030513 1.0000000
```

To work with the data more, first we can turn the grouse matrix into a tibble called **grouse_df**. Then, we will modify it to a long format to allow us to visualize it with `ggplot()`:

```
grouse_df <- cbind(as_tibble(prop.table(grouse)),
                   State = c("Michigan","Minnesota","Wisconsin"))

grouse_df <- grouse_df %>%
  pivot_longer(!State, names_to = "Outcome", values_to = "Proportion")

grouse_df
```

```
## # A tibble: 6 x 3
##    State      Outcome               Proportion
##    <chr>      <chr>                      <dbl>
## 1 Michigan   West Nile Negative         0.257
## 2 Michigan   West Nile Positive         0.0388
## 3 Minnesota  West Nile Negative         0.331
## 4 Minnesota  West Nile Positive         0.0472
## 5 Wisconsin  West Nile Negative         0.232
## 6 Wisconsin  West Nile Positive         0.0943
```

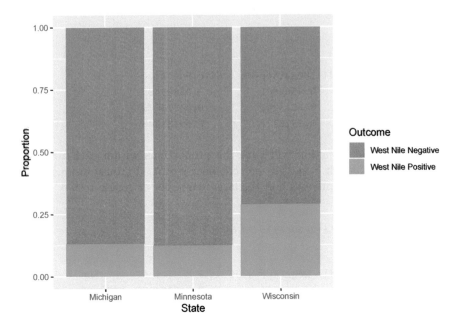

FIGURE 6.2 The distribution of West Nile infections in ruffed grouse across three states.

The following graph plots a stacked bar graph showing the proportion of West Nile infections within each state. We observe that Wisconsin has a higher positive infection rate of West Nile compared to Michigan and Minnesota:

```
ggplot(grouse_df, aes(x = State, y = Proportion, fill = Outcome)) +
  geom_bar(stat = "identity", position = 'fill')
```

If we were to calculate the χ^2 statistic by hand, we would start by determining the expected cell counts. We will begin with grouse in Michigan that tested negative for West Nile. Recall that 213 grouse were from Michigan, 591 grouse tested negative for the virus, and 721 total grouse were sampled:

$$\text{expected}_{\text{MI, Neg}} = \frac{\text{row total} \cdot \text{column total}}{n} = \frac{213 \cdot 591}{721} = 174.6$$

So, we would expect 174.6 grouse in Michigan would test negative for West Nile. In the calculation of the χ^2 statistic we will compare this value to the observed number of grouse (185). We'll then determine the expected number of grouse in Michigan that tested positive for West Nile:

TABLE 6.1 Chi-squared table of the grouse data.

State, Outcome	Observed	Expected	(Obs- Exp)^2 / Exp
MI, Neg	185	174.6	0.62
MI, Pos	28	38.4	2.82
MN, Neg	239	223.8	1.03
MN, Pos	34	49.2	4.70
WI, Neg	167	192.6	3.40
WI, Pos	68	42.4	15.46
SUM	721	721.0	28.02

$$\text{expected}_{\text{MI, Pos}} = \frac{213 \cdot 130}{721} = 38.4$$

Then we'll calculate the expected cell counts for all other cells in the table:

$$\text{expected}_{\text{MN, Neg}} = \frac{273 \cdot 591}{721} = 223.8$$

$$\text{expected}_{\text{MN, Pos}} = \frac{273 \cdot 130}{721} = 49.2$$

$$\text{expected}_{\text{WI, Neg}} = \frac{235 \cdot 591}{721} = 192.6$$

$$\text{expected}_{\text{WI, Pos}} = \frac{235 \cdot 130}{721} = 42.4$$

Table 6.1 summarizes the observed and expected counts of the grouse data. The last column calculates the squared difference divided by the expected count and sums them to obtain the χ^2 statistic.

Note that the greatest value in the table of outcomes (15.46) is for grouse in Wisconsin that tested positive for the virus. This is due to the greater number of observed grouse with the virus (68) compared to the expected number (42.4). The greater proportion of infections in Wisconsin can also be seen in Figure 6.2. Minnesota saw a lower number of observed grouse with the virus (34) compared to the expected number (49.2). Together, these outcomes provide large values that go into the calculation of the χ^2 statistic.

The degrees of freedom for the grouse data are $(3 - 1)(2 - 1) = 2$. With a $\chi^2(2)$ distribution and a large value of the statistic (28.02), we will likely have evidence to support H_A and conclude that there is a relationship between West Nile infection and the state where grouse were sampled.

6.4   The chi-squared test in R

We were introduced to the `chisq.test()` function in Chapter 5 when we were dealing with data in a 2x2 table. The function is flexible to include any number of outcomes. For example, the grouse data are stored as a 3x2 table. If the data are stored in a matrix format, the `chisq.test()` function can be easily applied to run a chi-squared test at the $\alpha = 0.05$ level:

```
chisq.test(grouse)
```

```
##
##   Pearson's Chi-squared test
##
## data:  grouse
## X-squared = 28.094, df = 2, p-value = 7.935e-07
```

Our manual calculation of the χ^2 statistic (28.02) produced a similar value to that provided by the function (28.094). Due to the small p-value (7.935e-07), we indeed have evidence to support H_A and conclude that there is a relationship between West Nile infection and the state where grouse were sampled.

6.4.1   Exercises

6.1 Determine the number of degrees of freedom for conducting a chi-squared test for the following case studies:

    a. A survey that asked participants their generation (Gen Z, Millennial, Gen X, or Boomer) and whether they have taken personal actions to address climate change in the last year (yes or no).

    b. A survey that asked participants their political party (conservative or liberal) and whether they believe that governments provide adequate funding to address climate change (yes or no).

    c. A survey that asked participants where they live (urban, rural, or suburban) and whether they believe governments should provide more, less, or the same amount of funding to address climate change.

6.2 Adedayo et al. (2010) investigated how rural women use forest resources in different regions in Nigeria. They surveyed women in three different ethnic groups according to different levels of income from forest resources, grouped in increments of 10%.

Run the code below to reproduce the data from Adedayo et al. (2010; their Table 2):

```
forest.resources <- matrix(c(2, 7, 22, 36, 10, 3,
                             5, 4, 20, 37, 8, 6,
                             0, 3, 28, 35, 10, 4),
                           ncol = 6, byrow = T)
colnames(forest.resources) <- c("<30%", "30-40%", "41-50%",
                                "51-60%", "61-70%", ">70%")
rownames(forest.resources) <- c("Yoruba", "Nupe", "Berom")
```

   a. How many degrees of freedom are associated with a chi-squared test for these data?
   b. Run code in R that tests the null hypothesis that the proportion of respondents' income from forest resources is independent from ethnic group at the $\alpha = 0.05$ level. State the results of the test in a sentence or two.

6.3. Run a similar chi-square test as above, but add the `simulate.p.value = TRUE` statement to the test. Research what this statement does by looking up the documentation for the `chisq.test()` function. How do the results of your test differ?

6.5   Summary

In the last chapter we learned about hypothesis tests for proportions. This was straightforward because we could analyze the data for two outcomes and two groups. Two-way tables extend this analysis by quantifying differences for any number of groups. Data that can be organized into two-way tables are widespread in natural resources and easy to visualize and interpret.

The chi-squared test and the statistic it provides measures how far the true counts are in a group/outcome and compares it to what we expect the number of counts to be. In doing this, hypotheses in chi-squared tests are set up to infer whether the groups are independent (the null hypothesis) or if there is a relationship between the groups (the alternative hypothesis). Base functions in R such as `prop.table()` and `addmargins()` allow you to explore and visualize two-way tables while the `chisq.test()` function performs chi-squared tests.

6.6   Reference

Adedayo, A.G., M.B. Oyun, and O. Kadeba. 2010. Access of rural women to forest resources and its impact on rural household welfare in North Central Nigeria. *Forest Policy and Economics* 12(6): 439–450.

7

Sample size and statistical power

7.1 Introduction

"Every second of every day, our senses bring in way too much data than we can possibly process in our brains." This was stated by entrepreneur Peter Diamandis about the role of so-called "big data" in the 21st century. We can spend our entire careers collecting data. (And some people do.) How do we know when we have collected enough data to answer our question of interest? And how confident can we be in the sample we've taken to obtain a reliable estimate of the population of interest?

This chapter will provide details for determining the appropriate number of samples to collect. We will revisit concepts from hypothesis testing as it relates to understanding the power of a statistical test.

7.2 How much data should I collect?

In natural resources, data collection costs time, effort, and money. It is rare to have the resources available to measure all of the units in a population, or a **complete enumeration**. Instead, collecting the appropriate number of samples will help you be efficient with your budget and simultaneously provide reliable data to make decisions. A plan to collect the appropriate number of samples should be designed only after knowing the budget available for the task and the desired precision from our sample to inform what we want to know about the population.

7.2.1 What makes a sample good?

Before we determine how many samples to collect, it is worth discussing what makes a sample good. In natural resources, a good sample has the following characteristics:

- **It is unbiased.** This means that the sample serves as a useful approximation of the population. **Sample bias** would result if our expected results differ from the true value being estimated.
- **It is efficient.** This means that it is easy to obtain the data and the data are representative of the population.
- **It is flexible.** For example, collecting data on wildlife is challenging because it may mean capturing and releasing live animals. Measuring trees in remote areas can take considerable time and effort to travel to the study site. Flexible samples allow us to address multiple objectives, particularly in cases when the data are difficult to collect.

The goal of sampling is to obtain data that are reliable and will allow you to make inference about a population. A good sample will collect the appropriate number of observations to meet the objectives of the question being asked.

7.2.2 Determining sample size

Determining the appropriate sample size (n) requires objective and subjective assessments. Objective assessments include determining the variability of the sample, e.g., by calculating its standard deviation. Subjective assessments include specifying the desired confidence level and allowable error of the variable of interest. By having this information prior to the start of a study, a reliable n can be obtained.

To start, the standard deviation of the variable of interest needs to be obtained to represent its variability. While this may seem challenging because the data have not been collected yet, with minimal effort you can determine the standard deviation using one or more of the following strategies:

- **Conduct a pilot study.** Consider spending a limited amount of time collecting a few samples to determine the standard deviation for the variable of interest.
- **Consult with experts and colleagues.** Ask others that may have collected data on your variable of interest and inquire about the variability of their data.
- **Research historical data.** In natural resources, many species and their populations have been measured and cataloged. Perform a literature review and look for tables and statistical results that report standard deviations

for your variable. Consult your organization's records for historical data
that may have been collected on your variable.

- **Calculate the standard deviation with a "quick and dirty" approximation.** From the empirical rule, we know that nearly all of the data will
be found within four standard deviations of the mean (assuming the data
are distributed normally). A "quick and dirty" approximation of the standard deviation is $\sigma \approx range/4$. If you have an estimate of the minimum and
maximum values for your variable, they can be used to approximate the
standard deviation.

Along with the mean, the standard deviation will be used to determine the
coefficient of variation (CV) of the variable of interest.

When we talk about confidence, we will also need a value t from the t-
distribution to represent the desired confidence interval. To determine the
sample size needed at a probability level of 0.90, we can use the qt() function
with an infinite number of degrees of freedom (df = Inf), reflecting the large
number of observations in the population. We specify the quantile of 0.95 to
incorporate 5% of the area on the upper and lower values of the curve:

```
qt(0.95, df = Inf)
```

```
## [1] 1.644854
```

So, the t-value for a 90% confidence interval is 1.64.

The final component needed to determine sample size is a subjective assessment of allowable error. The allowable error, represented by A, is expressed
as the percent of the mean. For example, A can be set to 10 to estimate the
number of samples required to estimate the mean of the population to be
within ±10%.

The sample size from a simple random sample can be calculated as:

$$n = \left[\frac{(t)(CV)}{A}\right]^2$$

where t, CV, and A represent the t value, coefficient of variation, and allowable
error, respectively.

For example, consider that fisheries biologists want to determine the appropriate number of yellow perch weights to measure. The mean and standard
deviation of perch weight is 260 and 101 grams, respectively. If we want to

FIGURE 7.1 Yellow perch, a fish native to North America. Photo: US Fish and Wildlife Service.

determine the number of samples required to estimate a population mean to within $\pm 10\%$ at a probability level of 0.90, we would calculate

$$n = \left[\frac{(1.64)[(101/260)*100]}{10} \right]^2 = 40.6$$

After rounding up, we would need to weigh 41 perch to reach the desired precision. In R, the following function num_samples() performs the calculation using the ceiling() function to round up the sample size:

```
num_samples <- function(t, CV, A){
  n = ((t * CV) / A)**2
  return(ceiling(n))
}

num_samples(t = qt(0.95, df = Inf),
            CV = (101/260)*100,
            A = 10)
```

```
## [1] 41
```

You can quickly see that by changing the subjective measures of the calculation (e.g., the *t*-value associated with the level of confidence), different sample sizes will result. The appropriate number of samples will be based on your desired level of precision and available resources to collect the data. (A series

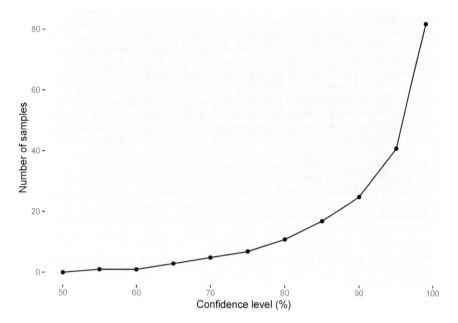

FIGURE 7.2 Changes in the number of samples of yellow perch weights needed to maintain a desired level of precision with an allowable error of 10%.

of upcoming exercises will investigate how changing the coefficient of variation and the allowable error A results in different sample sizes.)

7.2.3 Exercises

7.1 Write a function that calculates a "quick and dirty" approximation of the standard deviation as mentioned in this chapter, where $\sigma \approx range/4$. Use it to calculate approximate standard deviations for the following variables:

a. A population of perch weights with a maximum and minimum of 350 and 60 grams, respectively.
b. The ages of white-tailed deer of 2, 3, 3, 4, 6 and 7 years.
c. The height (HT) of cedar elm trees (in feet) from the **elm** data set.

7.2 Consider a quantitative variable that interests you (e.g., the weights of perch; the heights of cedar elm trees). Perform an internet search to find a scientific article or publication that reports a mean and standard deviation for this value. Use R code to calculate the coefficient of variation for this variable of interest.

7.3 A strawberry farmer collects some data on yields from different fields on her farm. She collects data from five fields and records yields of 48,000, 52,000,

FIGURE 7.3 Strawberries. Image: domdomegg/Wikimedia Commons.

65,000, 58,000, and 56,000 pounds per acre. Use the `num_samples()` function to determine the appropriate number of samples to collect under the following scenarios.

    a. Find the sample size necessary to estimate a population mean with 95% confidence with an allowable error of 5%.

    b. Find the sample size necessary to estimate a population mean with 99% confidence with an allowable error of 5%.

7.4 Foresters measure the diameters of logs to sell to a sawmill. Data are a sample from a population of hundreds of logs from a recent timber harvest. They calculate a mean log diameter of 22.6 inches and a standard deviation of 8.5 inches. Use the `num_samples()` function to determine the appropriate number of samples under the following scenarios.

    a. Find the sample size necessary to estimate a population mean with 90% confidence with an allowable error of 15%.

    b. How many more logs need to be sampled if the coefficient of variation of the logs is doubled?

    c. How many fewer logs need to be sampled if the coefficient of variation of the logs is cut in half?

    d. How many fewer logs need to be sampled if the allowable error of the logs is doubled?

    e. How many more logs need to be sampled if the allowable error of the logs is cut in half?

7.3 Statistical power

We learned in Chapter 4 that Type II error occurs when we fail to reject the null hypothesis when it is in fact false, denoted by β. The probability of correctly rejecting the null hypothesis when it is indeed false is $1 - \beta$, also known as the **statistical power** of a hypothesis test.

7.3.1 Defining statistical power

To understand statistical power, it is worth revisiting the concepts of p-values. A **p-value** is the probability under a specified statistical model that a statistical summary of the data would be equal to or more extreme than its observed value. If the p-value is smaller than some pre-defined significance level termed α (typically set to 0.05), the result is said to be statistically significant.

The probability of obtaining a statistically significant result is called the **statistical power** of the test. The power of a test relies on the size of the effect and the sample size.

Statistical power ranges from 0 to 1. As power increases, the probability of making a Type II error decreases. In natural resource sciences, specifying a statistical power of 0.80 is common.

7.3.2 Calculating the power of statistical tests

R's power.t.test() function determines the power of a statistical hypothesis test. You will need to know several things prior to calculating the power:

- the null and alternative hypotheses for your test,
- whether you're conducting a one- or a two-sided test, paired test, or test of proportions,
- the sample size,
- the sample mean and standard deviation,
- the difference between your hypothesized mean value and your sample mean (represented by delta =), and
- the significance level α.

For example, we can calculate the power of a two-sided one-sample t-test on the weights of yellow perch. Consider a sample of 150 perch with $\bar{y} = 260$ and $s = 101$ grams. The null hypothesis H_0 is that the true mean is equal to 240 grams and the alternative hypothesis H_1 is that the true mean is not equal to 240 grams:

```
power.t.test(n = 150,
             delta = 260 - 240,
             sd = 101,
             sig.level = 0.05,
             type = "one.sample",
             alternative = "two.sided")
```

```
##
##      One-sample t test power calculation
##
##                 n = 150
##             delta = 20
##                sd = 101
##         sig.level = 0.05
##             power = 0.6735067
##       alternative = two.sided
```

Note the output provides you the power of the *t*-test. A greater power indicates that we have a much better ability to detect a difference if one exists. The results from this test indicate we have a 67% chance of finding a difference if one exists in the population.

We can also use the `power.t.test()` function to perform a two-sample *t*-test on the weights of yellow perch. Consider two samples of perch with 150 observations each with $\bar{y}_1 = 240$ $\bar{y}_2 = 270$. The null hypothesis H_0 is that the two means are equal and the alternative hypothesis H_1 is that the true means are not equal:

```
power.t.test(n = 150,
             delta = 240 - 270,
             sd = 101,
             sig.level = 0.05,
             type = "two.sample",
             alternative = "two.sided")
```

```
##
##      Two-sample t test power calculation
##
##                 n = 150
##             delta = 30
```

```
##                   sd = 101
##        sig.level = 0.05
##            power = 0.7271084
##      alternative = two.sided
##
## NOTE: n is number in *each* group
```

The results from this two-sample test indicate we have a 73% chance of finding a difference if one exists in the population. You may change the argument to alternative = "one.sided" to find the power of a one-sided test.

The power.t.test() function can also be used for calculating the power of a paired *t*-test by changing type = "paired". For proportions, the power.prop.test() function can calculate the power of two-sample tests.

We can also use the power.t.test() function to ask how many samples are needed to achieve a desired level of power. If we state n = NULL in the code and specify power = 0.80, the function will tell us how many samples would be needed to achieve 0.80 power.

Using the two-sided one-sample *t*-tests on yellow perch weights described above, the following code will find the number of samples needed to carry out a two-sided one-sample *t*-test with 99% confidence. The null hypothesis is that yellow perch weight is equal to 240 grams and an alternative hypothesis that weight is not equal to 240 grams:

```
power.t.test(n = NULL,
             delta = 260 - 240,
             sd = 101,
             power = 0.80,
             sig.level = 0.01,
             type = "one.sample",
             alternative = "two.sided")
```

```
##
##      One-sample t test power calculation
##
##                n = 301.1659
##            delta = 20
##               sd = 101
##        sig.level = 0.01
##            power = 0.8
##      alternative = two.sided
```

We would need to collect 301.1659, or 302 samples of yellow perch weight to achieve the desired power of 0.80.

We can perform a similar calculation on proportions with the power.prop.test() function. For example, consider two technicians that inspect facilities for hazardous waste compliance. There is a probability of 0.21 and 0.32 that they will issue a violation on each visit. A 0.75 power calculation with a significance level $\alpha = 0.05$ would be:

```
power.prop.test(n = NULL,
                power = 0.75,
                p1 = 0.21,
                p2 = 0.32,
                sig.level = 0.05)
```

```
##
##      Two-sample comparison of proportions power calculation
##
##                   n = 222.548
##                  p1 = 0.21
##                  p2 = 0.32
##           sig.level = 0.05
##               power = 0.75
##         alternative = two.sided
##
## NOTE: n is number in *each* group
```

We would need to collect 222.548, or 223 samples from each technician to achieve the desired power of 0.75.

7.3.3 Exercises

7.5 For the following questions, use the **elm** data set and conduct the analyses with a significance level of $\alpha = 0.10$.

a. Calculate the power of a two-sided one-sample *t*-test on the height (HT) of elm trees. The null hypothesis of the test is that the true mean of all trees is equal to 30 feet and the alternative hypothesis that the true mean is not equal to 30 feet.

b. Calculate the power of a two-sided one-sample *t*-test on the height (HT) of all co-dominant elm trees. The null hypothesis of the test is that the true mean of co-dominant elm trees is equal to 34 feet and the alternative hypothesis that the true mean is not equal to 34 feet.

c. Find the number of samples of elm trees needed to achieve 0.85 power based on the hypothesis test outlined in part a.

d. Find the number of samples of co-dominant elm trees needed to achieve 0.70 power based on the hypothesis test outlined in part b.

7.6 Using the hypothesis test in question 7.5a, find the number of samples of elm trees needed to achieve different levels of power ranging in ten percent increments from 0.50 to 0.90. Make a plot using ggplot() that shows the influence of power on the sample size.

7.7 The following proportions of ruffed grouse showed antibodies consistent with West Nile virus in different US states: Michigan (0.13), Minnesota (0.12), and Wisconsin (0.29). Calculate the number of grouse samples needed in each state to achieve a power of 0.80 at a significance level $\alpha = 0.05$ using the following two-sided tests:

a. Michigan and Minnesota
b. Michigan and Wisconsin
c. Minnesota and Wisconsin
d. Why do the sample sizes required in the previous tests differ?

7.4 Summary

Determining the appropriate number of samples to collect is a critical part of natural resources science and management. Sampling efforts in natural resources need to be unbiased, flexible, and efficient. Understanding our variables of interest before collecting a thorough sample will assist us by knowing the variability we should expect in the population. All efforts that determine sample size rely on both objective (e.g., standard deviation) and subjective (e.g., significance level) assessments.

When combined with a planned hypothesis test, the power of a statistical test allows us to minimize Type II error, or failing to reject the null hypothesis when it is in fact false. Tests of power can be conducted with the familiar hypothesis tests we learned in Chapter 4, including one- and two-sample and paired t-tests. A thorough understanding of the statistical concepts of sample size and power allows you to design an efficient sampling scheme, saving you time and energy by not oversampling.

8

Linear regression

8.1 Introduction

Regression is one of the most useful concepts in statistics because it involves numerical prediction. Regression predicts something we do not know, where the answer is a number. We often collect data on things that are easy to measure and call these the **independent** or **explanatory** variables. Regression is used to tell us about the things that are often hard to measure, termed the **dependent** or **response** variable.

As an example, consider you go fishing. It is relatively easy to measure the length of a fish when you bring it out of the water, but it is more difficult to measure how much it weighs. Using regression, we might develop a statistical model to predict the weight of the fish (the dependent variable) using its length (the independent variable). This would be an example of **simple linear regression**, or using a single independent variable in a regression to predict the dependent variable.

Regression allows us to make estimations and inference from one variable to another. When viewing the data as a scatter plot, if a linear pattern is evident, we can describe the relationships between variables by fitting a straight line through the points. Regression techniques will allow us to draw a line that comes as close as possible to the points.

8.2 Correlation

While regression allows us to make predictions, **correlation** measures the degree of association between two variables. Correlation coefficients range from -1 to 1, with negative correlations less than 0 and positive correlations greater than 0. Coefficients that are "extreme," i.e., they are closer to -1 or 1, are considered strongly correlated. A correlation coefficient near 0 indicates no correlation between two variables.

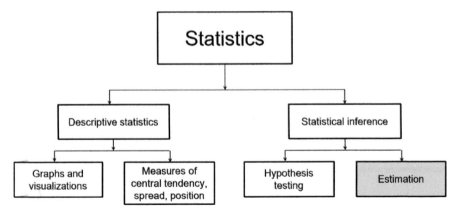

FIGURE 8.1 The remainder of this book focuses largely on estimation, a core component of statistical inference.

The Pearson correlation coefficient, denoted r, is appropriate if two variables x and y are linearly related. We can use all observations in our data from $i=1$ to n and employ the following formula to determine the coefficient:

$$r = \frac{\sum_{i=1}^{n}(x_i - \bar{x})(y_i - \bar{y})}{\sqrt{\sum_{i=1}^{n}(x_i - \bar{x})^2 \sum_{i=1}^{n}(y_i - \bar{y})^2}}$$

The following graphs show examples of correlations between three sets of variables:

- a store's price for a gallon of milk and their corresponding monthly sales,
- the length and width of sepals measured on four different species of iris flowers, and
- the number of chirps per second a ground cricket makes based on the air temperature.

In R, the cor() function computes the correlation between two variables x and y. We can apply it to determine the correlation between the length and width of sepal measurements contained in the **iris** data set:

```
iris <- tibble(iris)

cor(iris$Sepal.Length, iris$Sepal.Width)
```

```
## [1] -0.1175698
```

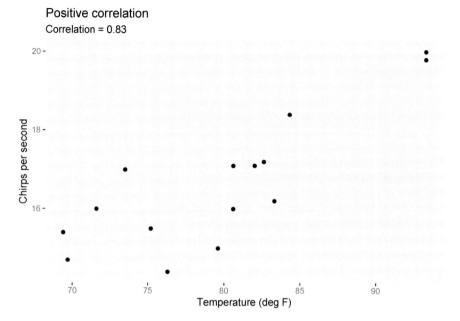

FIGURE 8.2 Correlation between temperature and the number of chirps a striped ground cricket makes.

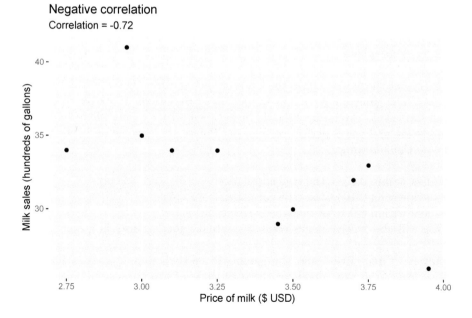

FIGURE 8.3 Correlation between milk price and sales.

The coefficient of 0.1176 indicates weak correlation between sepal length and width. The `cor.test()` function also computes the correlation coefficient and provides additional output. This function tests for the association between two variables by running a hypothesis test and providing a corresponding confidence interval:

```
cor.test(iris$Sepal.Length, iris$Sepal.Width)
```

```
##
##  Pearson's product-moment correlation
##
## data:  iris$Sepal.Length and iris$Sepal.Width
## t = -1.4403, df = 148, p-value = 0.1519
## alternative hypothesis: true correlation is not equal to 0
## 95 percent confidence interval:
##  -0.27269325  0.04351158
## sample estimates:
##         cor
## -0.1175698
```

In the case of the sepal measurements, the value 0 is included within the 95% confidence interval, indicating we would accept the null hypothesis that the correlation is equal to zero.

8.2.1 Exercises

8.1 Try out the game "Guess the Correlation" at http://guessthecorrelatio n.com/. The game analyzes how we perceive correlations in scatter plots. For each scatter plot, your guess should be between zero and one, where zero is no correlation and one is perfect correlation. Points are awarded and deducted if you do the following:

- Guess within 0.05 of the true correlation: +1 life and +5 coins
- Guess within 0.10 of the true correlation: +1 coin
- Guess within >0.10 of the true correlation: -1 life
- You will also receive bonus coins if you make good guesses in a row.

You have three "lives" and the game will end when your guesses in three consecutive tries are outside of the 0.10 value of the true correlation. After your game ends, what was your mean error and how many total points did you receive?

8.2. Use the `filter()` function to subset the iris data into two data sets that include the *setosa* and *virginica* species, separately. Run the `cor()` function to find the Pearson correlation coefficient between sepal length and sepal width for each data set. How would you describe the strengths of the correlations between the variables. What values do you find that are different from what was observed when the `cor()` function was presented in the text?

8.3 Run the `cor.test()` function to find the Pearson correlation coefficient between petal width and sepal width for the *Iris virginica* observations. If the null hypothesis is that the true correlation between the two variables is equal to 0, do you reject or fail to reject the test?

8.3 Concepts that build to regression

8.3.1 How hot is it?

Many international researchers are well aware of the process for converting between Fahrenheit and Celsius scales of temperature. We'll create a small data set called **degs** that has a range of temperatures that contain the equivalent values in both units:

```
degs <- tribble(
  ~degC, ~degF,
  0, 32,
  10, 50,
  20, 68,
  30, 86,
  40, 104
)
```

In our example, we wish to predict the degrees Fahrenheit (the dependent variable) based on the degrees Celsius (the independent variable). In regression, dependent variables are traditionally plotted on the y-axis and independent variables on the x-axis:

```
ggplot(degs, aes (x = degC, y = degF)) +
  geom_point() +
  geom_line()
```

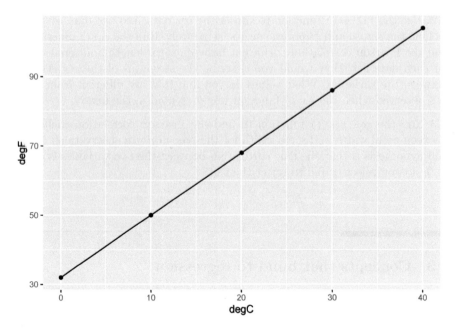

FIGURE 8.4 The (perfect) relationship between degrees Fahrenheit and Celsius.

Clearly, the relationship between degrees Fahrenheit and Celsius follows a linear pattern. To find the attributes of the fitted line, consider the equation $degF = \beta_0 + \beta_1 degC$, where β_0 and β_1 represent the y-intercept and slope of the regression equation, respectively. The y-intercept represents the value of y when the value of x is 0. To determine the y-intercept in our example, we can find the values in our equation knowing that when it's 0 degrees Celsius, its 32 degrees Fahrenheit:

$$32 = \beta_0 + \beta_1(0)$$

$$\beta_0 = 32$$

The slope can be determined by quantifying how much the line increases for each unit increase in the independent variable, or calculating the "rise over run." In our example, we know that when it's 10 degrees Celsius, its 50 degrees Fahrenheit. This can be used to determine the slope:

$$50 = 32 + \beta_1(10)$$

$$18 = \beta_1(10)$$

$$\beta_1 = 18/10 = 1.8$$

So, we can determine the temperature in degrees Fahrenheit if we know the temperature in degrees Celsius using the formula $degF = 32 + 1.8(degC)$. This formula is widely used in the weather apps on your smartphone and in online conversion tools. While there is no doubt a linear relationship exists between Celsius and Fahrenheit, more specifically we can say that there is a *perfect relationship* between the two scales of temperature: when it is 30 degrees Celsius, it is *exactly* 86 degrees Fahrenheit.

8.3.2 How hot will it be today?

Imagine you're outdoors with eight friends and ask each of them what they think the current air temperature is. People perceive temperatures differently, and other weather conditions such as wind, humidity, and cloud cover can influence a person's guess. You glance at your smartphone and note that the actual temperature is 28 degrees Celsius, but your friends have different responses, recorded in the **temp_guess** data set:

```
temp_guess <- tribble(
  ~FriendID, ~degC,
  1, 31,
  2, 25,
  3, 26,
  4, 27,
  5, 33,
  6, 29,
  7, 28,
  8, 30
)
```

```
ggplot(temp_guess, aes (x = degC, y = 1)) +
  geom_boxplot()
```

These true values and "predicted" values get us thinking about the error (or noise) that is inherent to all predictions. Your friends have made predictions with errors which can be characterized by the following equation:

$$\hat{y} = y + \varepsilon_i$$

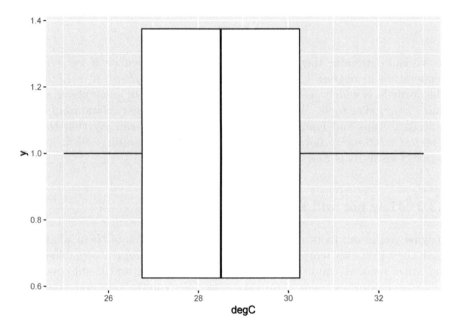

FIGURE 8.5 Box plot showing the distribution of temperature guesses around the true temperature of 28 degrees.

where $\hat{y}$ is the temperature guess, y is the true temperature, and ε_i is the measurement error. In regression, ε_i is the **residual**, or a measure of the difference between an observed value of the response variable and the prediction. We can create a variable called resid which calculates the residual values for temperature guesses and graphs them:

```
temp_guess  %>%
  mutate(resid = 28 - degC) %>%
  ggplot(aes(x = FriendID, y = resid)) +
  geom_point()
```

One person (Friend 7) correctly guessed the temperature, indicated by a residual of 0. Three friends predicted more than the true value of 28 degrees Celsius and four friends predicted less than the true value. This example can be considered a *location problem*, with guesses surrounding the true value.

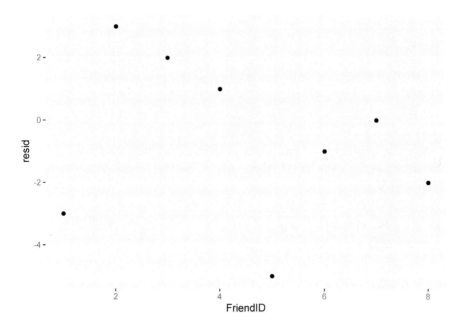

FIGURE 8.6 Residuals from the temperature guesses of friends.

8.3.3 How much heat can you tolerate?

Assume you present the same eight friends with a scenario: you ask them how much time they are willing to spend outdoors (in number of hours) doing physical activity when it is 20 degrees Celsius (68 degrees Fahrenheit). You ask the same question to your friends but inquire about how much time they are willing to do an activity at 40 degrees Celsius (104 degrees Fahrenheit).

```
outdoors <- tribble(
  ~FriendID, ~degC, ~hours,
  1, 20, 3.5,
  2, 20, 3.5,
  3, 20, 2,
  4, 20, 4,
  5, 20, 5,
  6, 20, 2.5,
  7, 20, 4,
  8, 20, 3,
  1, 40, 1,
  2, 40, 2,
  3, 40, 0.5,
```

```
  4, 40, 1,
  5, 40, 3,
  6, 40, 1,
  7, 40, 1,
  8, 40, 2
)
```

To view the responses, we can make a scatter plot that shows the number
of hours spent for each friend. Using `col = factor(FriendID)` provides the
scatter plot:

```
ggplot(outdoors, aes(x = degC, y = hours, col = factor(FriendID))) +
  geom_point()
```

Data indicate that your friends would spend more time outdoors at 20 degrees
rather than 40 degrees Celsius. But what if the temperature was 30 degrees
Celsius? You could visualize that the mean value across your eight friends

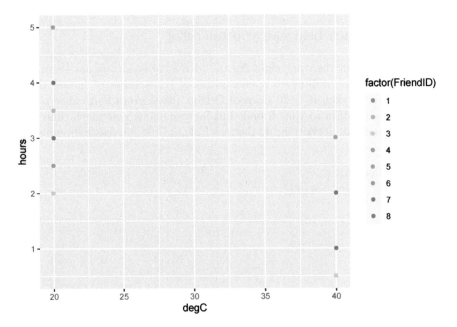

FIGURE 8.7 Number of hours people are willing to spend outdoors doing
physical activity at 20 and 40 degrees Celsius.

would be somewhere between 20 and 40 degrees Celsius. This is a slightly more difficult problem that requires interpolation.

Consider the regression equation $\hat{y} = \beta_0 + \beta_1(x) + \varepsilon_i$. We can find the values for β_0 and β_1 by first finding the mean values for the number of hours spent outdoors at 20 and 40 degrees Celsius:

$$\bar{y}_{20} = \frac{3.5 + 3.5 + \dots + 3}{8} = 3.44$$

$$\bar{y}_{40} = \frac{1 + 2 + \dots + 2}{8} = 1.44$$

In this case, we can use a centered regression equation to predict the number of hours spent outdoors:

$$\hat{y} = \beta_0 + \beta_1(x - \bar{x})$$

An obvious estimate when the air temperature is 30 degrees Celsius is the average of values calculated above, or

$$\beta_0 = \frac{3.44 + 1.44}{2} = 2.44$$

So, as the temperature goes from 30 to 40 degrees Celsius, the number of estimated hours spent outdoors goes from 2.44 to 3.44. So,

$$\beta_1 = \frac{2.44 - 3.44}{10} = -0.10$$

Hence, an equation predicting the number of estimated hours spent outdoors is

$$\hat{y} = 2.44 - 0.10(x - 30)$$

The benefit of this equation is that is can be applied to any temperature. For example, when it's 36 degrees Celsius, the number of estimated hours spent outdoors is $\hat{y} = 2.44 - 0.10(36 - 30) = 1.84$.

8.3.4 Exercises

8.4 Find the parameters for a regression equation that converts the distance measurement meters to feet. The **dist** data set has a range of measurements that contain equivalent values for both units:

```
dist <- tribble(
  ~dist_m, ~dist_ft,
  2, 6.56,
  5, 16.40,
  10, 32.80,
  20, 65.62
)
```

a. Using the **dist** data, create a scatter plot with lines using ggplot()
   that shows the measurements in feet on the y-axis (the dependent
   variable) and the measurements in meters on the x-axis (the independent variable).
b. On paper, find the parameters of the fitted line using the equation:
   dist_ft = beta_0 + beta_1(dist_m).
c. Write a function in R that uses the equation you developed. Apply
   it to determine the distance in feet for the following measurements
   in meters: 4.6, 7.7, and 15.6.

8.5 Christensen et al. (1996) reported on the relationship between coarse
woody debris along the shorelines of 16 different lakes in the Great Lakes
region (their Table 1). They observed a positive relationship between the density of coarse woody debris (cwd_obs; measured in $m^2 km^{-1}$) and the density
of trees (TreeDensity; measured in number of trees per hectare). The **cwd**
data set is from the **stats4nr** package and contains the data:

```
library(stats4nr)
head(cwd)
```

```
## # A tibble: 6 x 2
##    cwd_obs TreeDensity
##      <dbl>       <dbl>
## 1      121        1270
## 2       41        1210
## 3      183        1800
## 4      130        1875
## 5      127        1300
## 6      134        2150
```

a. The authors developed a regression equation to predict the density
   of coarse woody debris. The equation was $CWD = -83.6 + 0.12 *$
   $TreeDensity$. Add a variable to the **cwd** data called cwd_pred that
   predicts the density of coarse woody debris for the 16 lakes. Add a

variable to the **cwd** data called `resid` that calculates the residual value for each observation.

b. Create a scatter plot using `ggplot()` of the residuals (y-axis) plotted against the observed CWD values (x-axis). Add the `stat_smooth()` statement to your plot to add a smoothed line through the plot. What do you notice about the trends in the residuals as the CWD values increase?

8.4 Linear regression models

Two primary linear regression techniques can be employed when a single independent variable is used to inform a dependent variable. These include the simple proportions model and the simple linear regression.

8.4.1 Simple proportions model

A **simple proportions model** makes the assumption that when the dependent variable is zero, the independent variable is also zero. This makes sense in many biological applications. You may have observed this in completing question 8.1, i.e., when the distance in meters is zero, the distance in feet is also zero. In this approach, imagine you "scale" the independent variable x by some value to make our predictions $\hat{y}$. In doing this, we multiply the independent variable by a proportion.

The simple proportion model is also termed a **no-intercept** model because the intercept β_0 is set to zero. The model is written as

$$y = \beta_1 x + \varepsilon$$

Predictions of y are proportional to x, but ε is not proportional to x. To determine the estimated slope $\hat{\beta}_1$ for the simple proportions model, we can employ the following formula:

$$\hat{\beta}_1 = \frac{\sum_{i=1}^{n} x_i y_i}{\sum_{i=1}^{n} x_i^2}$$

To see how R implements regression, we'll use the air quality data set in R that provides daily ozone measurements collected in New York City from May to September 1973. We'll name the data set **air** using the `tibble()` function:

FIGURE 8.8 Smog in New York City, May 1973. Image: US Environmental Protection Agency.

```
air <- tibble(airquality)
```

Two variables of interest are (1) Ozone, the mean ozone in parts per billion from 1300 to 1500 hours measured at Roosevelt Island and (2) Temp, the maximum daily temperature in degrees Fahrenheit at LaGuardia Airport:

We'll make a scatter plot using ggplot() that shows the amount of ozone as the response variable and temperature as the explanatory variable. Note you will see a warning message indicating the plot removed rows containing missing values. You can see that generally when temperature is less than 80 degrees F there is little variation in ozone, but when temperature is greater than 80 degrees F ozone increases with warmer temperatures:

```
ggplot(air, aes(x = Temp, y = Ozone)) +
  geom_point() +
  labs(y = "Ozone (ppb)", x = "Temperature (deg F)")
```

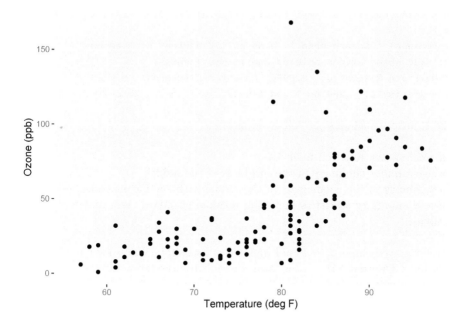

In R, the `lm()` function fits regression models. We will use it to model ozone (the response variable) using temperature (the independent variable). Response variables are added to the left of the ~ sign and independent variables to the right. Adding a *-1* after the independent variable will specify that we're interested in fitting a simple proportions model. We'll name the regression model object `lm.ozone` and use the `summary()` function to obtain the regression output:

```
lm.ozone <- lm(Ozone ~ Temp -1, data = air)
summary(lm.ozone)
```

```
##
## Call:
## lm(formula = Ozone ~ Temp - 1, data = air)
##
## Residuals:
##     Min     1Q Median     3Q    Max
## -38.47 -23.26 -12.46  15.15 121.96
##
## Coefficients:
##      Estimate Std. Error t value Pr(>|t|)
## Temp  0.56838    0.03498   16.25   <2e-16 ***
## ---
```

```
## Signif. codes:  0 '***' 0.001 '**' 0.01 '*' 0.05 '.' 0.1 ' ' 1
##
## Residual standard error: 29.55 on 115 degrees of freedom
##   (37 observations deleted due to missingness)
## Multiple R-squared:  0.6966, Adjusted R-squared:  0.6939
## F-statistic:   264 on 1 and 115 DF,  p-value: < 2.2e-16
```

The R output includes the following metrics:

- a recall of the function fit in `lm()`,
- a five-number summary of the residuals of the model,
- a summary of the coefficients of the model with significant codes, and
- model summary statistics including residual standard error and R-squared values.

The estimated slope $\hat{\beta}_1$ is found under the *Estimate* column next to the variable `Temp`. To predict the ozone level, we multiply the temperature by 0.56838. In other words, the formula can be written as $Ozone = 0.56838 * Temp$. As an example, if the temperature is 70 degrees, the model would predict an ozone level of $0.56838 * 70 = 39.8ppb$.

8.4.2 Simple linear regression

The simple linear regression model adds back the y-intercept value β_0 that we exclude in the simple proportions model:

$$y = \beta_0 + \beta_1 x + \varepsilon$$

When we predict y, we want the regression line to be as close as possible to the data points in the vertical direction. This means selecting the appropriate values of β_0 and β_1, bringing us to the concept of **least squares**. We will find the values for the intercept and slope that minimizes the residual sums of squares (SSR), calculated as:

$$SSR = \sum_{i=1}^{n} \varepsilon_i^2 = \sum_{i=1}^{n} (y_i - (\hat{\beta}_0 + \hat{\beta}_1 x))^2$$

The residuals ε_i can be visualized as the vertical distances between the points and the least-squares regression line:

If we were to manually calculate the values for the intercept and slope in a simple linear regression, we need to determine the mean and standard deviation of both x and y and make some other calculations. The slope $\hat{\beta}_1$ is calculated as:

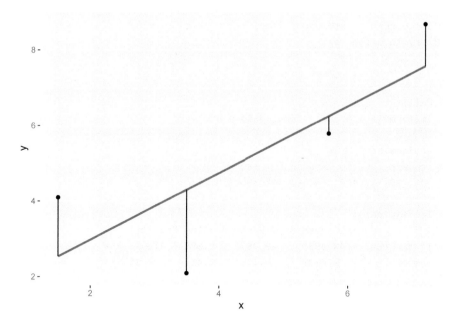

FIGURE 8.9 Regression line (blue) with observed values (black) and residual values representing the vertical distance between observed values and the regression line.

$$\hat{\beta}_1 = \frac{S_{xy}}{S_{xx}}$$

where

$$S_{xy} = \sum_{i=1}^{n}(x_i - \bar{x})(y_i - \bar{y})$$

$$S_{xx} = \sum_{i=1}^{n}(x_i - \bar{x})^2$$

You can read both S_{xy} and S_{xx} as the "sum of all x and y values" and the "sum of all squared x values." After calculating the slope, the intercept can be found as:

$$\hat{\beta}_0 = \bar{y} - \hat{\beta}_1\bar{x}$$

The lm() function in R is much simpler for a simple linear regression model because we can remove the -1 statement that we added for the simple proportions model. The regression model object lm.ozone.reg specifies the simple linear regression for the New York City ozone data:

```
lm.ozone.reg <- lm(Ozone ~ Temp, data = air)
summary(lm.ozone.reg)
```

```
##
## Call:
## lm(formula = Ozone ~ Temp, data = air)
##
## Residuals:
##     Min      1Q  Median      3Q     Max
## -40.729 -17.409  -0.587  11.306 118.271
##
## Coefficients:
##              Estimate Std. Error t value Pr(>|t|)
## (Intercept) -146.9955    18.2872  -8.038 9.37e-13 ***
## Temp           2.4287     0.2331  10.418  < 2e-16 ***
## ---
## Signif. codes:  0 '***' 0.001 '**' 0.01 '*' 0.05 '.' 0.1 ' ' 1
##
## Residual standard error: 23.71 on 114 degrees of freedom
##   (37 observations deleted due to missingness)
## Multiple R-squared:  0.4877, Adjusted R-squared:  0.4832
## F-statistic: 108.5 on 1 and 114 DF,  p-value: < 2.2e-16
```

Note that in the *Coefficients* section we now have an estimate for both the intercept and slope. The simple linear regression formula can be written as $Ozone = -146.9955 + 2.4287 * Temp$. If the temperature is 70 degrees, the model would predict an ozone level of $-146.9955 + 2.4287 * 70 = 23.0ppb$.

The slopes for the simple proportions and simple linear regression model result in starkly different values: 0.56838 and 2.4287. This also explains the differences noted between the prediction of ozone when the temperature is 70 degrees (39.8 versus 23.0 ppb). The positioning of where the regression line crosses the y-axis (i.e., the y-intercept) influences your predictions. For the ozone predictions, we can add linear regression lines for the simple proportions and simple linear regression model using the geom_abline() statement and adding the values for the slope and intercept:

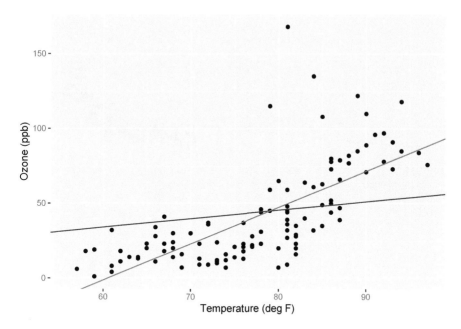

FIGURE 8.10 Regression lines for a simple proportions model (blue) and simple linear regression model (red) applied to the ozone data.

```
ggplot(air, aes(x = Temp, y = Ozone)) +
  geom_point() +
  labs(y = "Ozone (ppb)", x = "Temperature (deg F)") +
  geom_abline(intercept = 0, slope = 0.56838, col = "blue") +
  geom_abline(intercept = -146.9955, slope = 2.4287, col = "red")
```

Note the differences in slopes as they appear through the data. We can "zoom out" to see the intercept values and where they cross the *y*-axis. To do this we can alter the scale of the axes using `scale_x_continuous()` and `scale_y_continuous()` and change the minimum and maximum values in the `limits` statement:

```
ggplot(air, aes(x = Temp, y = Ozone)) +
  geom_point() +
  labs(y = "Ozone (ppb)", x = "Temperature (deg F)") +
  geom_abline(intercept = 0, slope = 0.56838, col = "blue") +
  geom_abline(intercept = -146.9955, slope = 2.4287, col = "red") +
  scale_x_continuous(limits = c(0, 100)) +
  scale_y_continuous(limits = c(-200, 200))
```

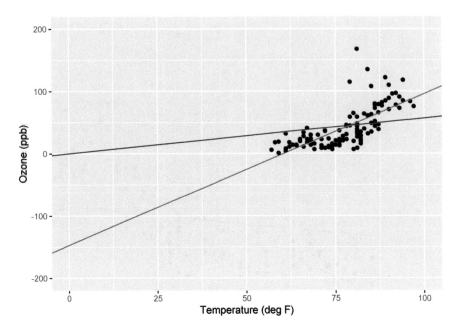

FIGURE 8.11 Regression lines showing y-intercept values for a simple proportions model (blue) and simple linear regression model (red) applied to the ozone data.

By looking at the data like this, we can see the impact of "forcing" the regression line through the origin of the simple proportions model. The range of temperature values is small, generally from 60 to 100 degrees Fahrenheit, and the simple linear regression line compensates for this by providing a negative value for the y-intercept.

8.4.3   Exercises

8.6 Replicate the model that Christensen et al. (1996) fit to the **cwd** data (see Question 8.5). Fit a simple linear regression using the lm() function with the density of coarse woody debris (cwd_obs) as the dependent variable and the density of trees (TreeDensity) as the independent variable. How do the values for the intercept and slope you find compare to what the original authors found?

8.7 Now, fit a simple proportions model to the **cwd** data. How does the value for the estimated slope $\hat{\beta}_1$ compare to the value you found in the simple linear regression?

8.8 Create a scatter plot showing the density of coarse woody debris and density of trees. Add a linear regression line to the scatter plot between the density of coarse woody debris and trees by typing `stat_smooth(method = "lm")` to your `ggplot()` code. This will present the simple linear regression line with 95% confidence bands.

8.5 Inference in regression

Up to this point we've been discussing regression in the context of being able to make predictions about variables of interest. We can use the principles of least squares to make inference on those variables, too. We can integrate hypothesis tests and other modes of inference into applications of linear regression. This includes performing hypothesis tests for the intercept and slope and calculating their confidence intervals. The components of linear regression can be viewed in an **analysis of variance table** which can tell us about the strength of the relationship between the independent and dependent variables.

A linear regression can estimate values for the intercept $\hat{\beta}_0$ and slope $\hat{\beta}_1$. But how do we know if they are any good? In other words, do the intercept and slope differ from 0? Null hypotheses can be specified for the slope as $H_0 : \hat{\beta}_1 = 0$ with the alternative hypothesis $H_A : \hat{\beta}_1 \neq 0$. The result of this hypothesis test will reveal if the regression line is flat or if it has a slope to it (either positive or negative).

We can conduct a simple linear regression between any two variables. This section will focus on how *good* the regression is and how we make inference from it.

8.5.1 Partitioning the variability in regression

We can partition the variability in a linear regression into its components. These include:

- the **regression sums of squares** $SS(Reg)$, the amount of variability explained by the regression model,
- the **residual sums of squares** $SS(Res)$, the error or amount of variability **not** explained by the regression model,

- the **total sums of squares** (TSS), the total amount of variability (i.e., $TSS = SS(Reg) + SS(Res)$).

If we have each observation of the dependent variable y_i, its estimated value from the regression $\hat{y}_i$, and its mean $\bar{y}$, we can calculate the components as

$$TSS = \sum_{i=1}^{n} (y_i - \bar{y})^2$$

$$SS(Reg) = \sum_{i=1}^{n} (\hat{y}_i - \bar{y})^2$$

$$SS(Res) = \sum_{i=1}^{n} (y_i - \hat{y}_i)^2$$

We can keep track of all of the components of regression in an **analysis of variance (ANOVA)** table. The table has rows for $SS(Reg)$, $SS(Res)$, and TSS. The degrees of freedom for $SS(Reg)$ are 1 because we're using one independent variable in a simple linear regression. The degrees of freedom for $SS(Res)$ are $n-2$ because we're estimating two regression coefficients $\hat{\beta}_0$ and $\hat{\beta}_1$. Just like the TSS can be found by summing the regression and residual component, so too can its degrees of freedom $(n-1)$.

Mean square values for the regression $(MS(Reg))$ and residual components $(MS(Res))$ can be calculated by dividing the sums of squares by the corresponding degrees of freedom. The value for $MS(Res)$ can be thought of as the "bounce" around the fitted regression line. In a "perfect" regression where you can draw a linear line between all points, $MS(Res)$ will be close to 0. With natural resources, we often work with data that are messy, so we expect larger values for $MS(Res)$.

In fact, $MS(Res)$ is really only meaningful for our calculations. If we take the square root of it, we obtain the **residual standard error**, the value provided in R output. In the ozone-temperature data, the residual standard error was 23.71 ppb of ozone. If temperature was not correlated with ozone, a higher residual standard error would result. Conversely, if temperature was more highly correlated with ozone, a lower residual standard error would result.

The value in the right-most column of the ANOVA table shows the F statistic, which is associated with a hypothesis test for the overall regression that tells you if the model has predictive capability. In simple linear regression, the F test examines a null hypothesis that the slope of your model $\hat{\beta}_1$ is equal to zero. In other words, regressions with larger F values provide evidence against the null hypothesis, i.e., a trend exists between your independent and dependent variable.

The ANOVA table and corresponding calculation can be viewed in a compact form:

TABLE 8.1 The analysis of variance (ANOVA) table in simple linear regression.

Source	Degrees of freedom	Sums of squares	Mean square	F
Regression	1	SS(Reg)	MS(Reg) = SS(Reg)/1	MS(Reg) / MS(Res)
Residual	n-2	SS(Res)	MS(Res) = SS(Res)/n-2	-
Total	n-1	TSS	-	-

TABLE 8.2 The analysis if variance (ANOVA) table for the ozone-temperature data.

Source	Degrees of freedom	Sums of squares	Mean square	F
Temp	1	61033	61033	108.53
Residual	114	64110	562	-
Total	115	125143	-	-

The anova() function in R produces the ANOVA table. Consider the ozone-temperature regression lm.ozone.reg from earlier in this chapter:

```
anova(lm.ozone.reg)
```

```
## Analysis of Variance Table
##
## Response: Ozone
##             Df Sum Sq Mean Sq F value    Pr(>F)
## Temp         1  61033   61033  108.53 < 2.2e-16 ***
## Residuals 114  64110     562
## ---
## Signif. codes:  0 '***' 0.001 '**' 0.01 '*' 0.05 '.' 0.1 ' ' 1
```

The row labeled *Residuals* corresponds to the residual sums of squares while the *Temp* row represents the regression sums of squares. The R output includes a corresponding p-value of the regression (Pr(>F)) which can then be compared to a specified level of significance to evaluate a hypothesis test on the regression (i.e., $\alpha = 0.05$). Note the total sums of squares is not shown in the R output, but to summarize we can add it to our table:

DATA ANALYSIS TIP: R attempts to make your life easier by also providing asterisks next to your variables indicating thresholds of significance. Aside from reporting these thresholds, it is always a good practice to report the actual *p*-value obtained from your statistical analysis. It is more transparent and allows your reader to better interpret your analysis.

As you can see in the figure, there is a healthy amount of "scatter" between the regression line and the ozone-temperature observations. This can be related to the sums of squares values from the regression. The ANOVA table indicates that the sums of squares for Temp is approximately the same as for the residuals (61,033 versus 64,110). More broadly, varying amounts of scatter around regression lines, along with the steepness of slopes, can be related to sums of squares values. Regressions with large values for $SS(Reg)$ result in steep slopes, while regressions with large values for $SS(Res)$ result in more scatter around the regression line:

After creating a scatter plot of your independent and dependent variables, you will have a solid understanding about the relative values of sums of squares and how they influence your regression.

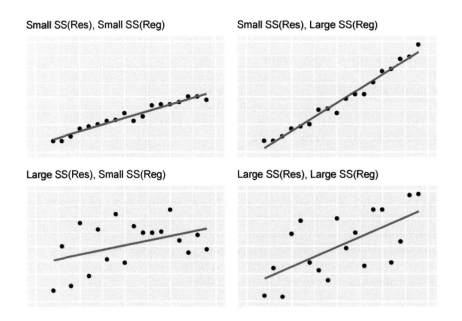

FIGURE 8.12 A visual representation of regression and residual sums of squares values as they relate to linear regression.

8.5.2 Coefficient of determination (R-squared)

The **coefficient of determination**, or more commonly termed, the R^2, measures the strength of the relationship between the dependent and independent variables. If this sounds similar to the Pearson's correlation coefficient, it's because it is. In simple linear regression, the R^2 value can be obtained by squaring r, the Pearson correlation coefficient. In doing this, R^2 always takes a value between 0 and 1. The R^2 is a widely used metric to assess the performance of a regression model.

Higher R^2 values indicate a greater portion of $SS(Reg)$ relative to TSS. As an example application, an R^2 of 0.75 means that 75% of the variability in the response variable is explained by the regression model. In other words,

- if $R^2 = 1$, all of the variability is explained by the regression model (i.e., large $SS(Reg)$) and
- if $R^2 = 0$, all of the variability is due to error (i.e., small $SS(Reg)$).

Values can be calculated by:

$$R^2 = \frac{SS(Reg)}{TSS} = 1 - \frac{SS(Res)}{TSS}$$

In R output, R^2 values are shown as `Multiple R-squared` values. In our example, 48.77% of the variability in the ozone is explained by temperature. An adjusted R^2 value, R^2_{adj}, can also be determined and is provided in R output. The R^2_{adj} value incorporates how many k independent variables are in your regression model and "penalizes" you for additional variables, as seen in the equation to determine the value:

$$R^2_{adj} = 1 - \left(\frac{(1 - R^2)(n - 1)}{n - k - 1} \right)$$

Values for R^2_{adj} will decrease as more variables are added to a regression model (e.g., in multiple regression, the focus of Chapter 9). Similarly, R^2_{adj} can never be greater than R^2. Interestingly, a mathematical proof can show that the value of R^2 will never decrease when new variables are added, even when those variables have no correlation with your response variable of interest.

In our example, the R^2_{adj} for the ozone regression is 48.32%, slightly less than the R^2 value. In simple linear regression, reporting R^2 is appropriate because $k = 1$. In applications when $k > 1$, R^2_{adj} is best to report.

8.5.3 Inference for regression coefficients

We learned earlier that the F statistic in regression examines whether or not the slope of $\hat{\beta}_1$ is equal to zero. Another way to examine the slope of our regression line is to make inference on $\hat{\beta}_1$ directly. Statistical tests are run on the regression coefficient associated with the slope parameter to examine the null hypothesis $H_0 : \hat{\beta}_1 = 0$. The t test statistic follows a t distribution with $n - 2$ degrees of freedom:

$$t = \frac{\hat{\beta}_1}{\sqrt{MS(Res)\left(\frac{1}{S_{xx}}\right)}}$$

Although it's less common in practice, you can also examine the intercept with a null hypothesis $H_0 : \hat{\beta}_0 = 0$ using the following test statistic:

$$t = \frac{\hat{\beta}_0}{\sqrt{MS(Res)\left(\frac{1}{n} + \frac{\bar{x}^2}{S_{xx}}\right)}}$$

You can also test whether or not $\hat{\beta}_0$ or $\hat{\beta}_1$ is different from a constant value other than 0 (termed $\beta_{0,0}$ and $\beta_{1,0}$). To do this, replace the numerator in the above equations with $\hat{\beta}_0 - \beta_{0,0}$ or $\hat{\beta}_1 - \beta_{1,0}$, respectively. In R, the summary() function provides inference for the regression coefficients in the *Std. Error*, *t value*, and *Pr(>|t|)* columns. Recall the lm.ozone.reg model:

```
summary(lm.ozone.reg)
```

```
##
## Call:
## lm(formula = Ozone ~ Temp, data = air)
##
## Residuals:
##      Min       1Q   Median       3Q      Max
## -40.729  -17.409   -0.587   11.306  118.271
##
## Coefficients:
##               Estimate Std. Error t value Pr(>|t|)
## (Intercept) -146.9955    18.2872  -8.038 9.37e-13 ***
## Temp           2.4287     0.2331  10.418  < 2e-16 ***
## ---
## Signif. codes:  0 '***' 0.001 '**' 0.01 '*' 0.05 '.' 0.1 ' ' 1
```

```
##
## Residual standard error: 23.71 on 114 degrees of freedom
##    (37 observations deleted due to missingness)
## Multiple R-squared:   0.4877, Adjusted R-squared:   0.4832
## F-statistic: 108.5 on 1 and 114 DF,   p-value: < 2.2e-16
```

The standard error of the regression coefficients are calculated in the denominators of the test statistics for $\hat{\beta}_0$ and $\hat{\beta}_1$ and the corresponding t-statistics are provided in R output. To gauge how "good" your regression coefficients are, relative to your regression estimates, values for standard errors should be small and t statistics large. Of course, R makes it easy for the analysts by providing the corresponding p-value for the intercept and slope in the rightmost column.

The **confidence interval** for the regression coefficient has a familiar form: estimate $\pm$ t x the standard error of estimate. After specifying a level of significance α, a $(1 - \alpha)100\%$ confidence interval for β_1 is

$$\hat{\beta}_1 = t_{n-2,\alpha/2}\sqrt{MS(Res)\left(\frac{1}{S_{xx}}\right)}$$

and a $(1 - \alpha)100\%$ confidence interval for β_0 is

$$\hat{\beta}_0 = t_{n-2,\alpha/2}\sqrt{MS(Res)\left(\frac{1}{n} + \frac{\bar{x}^2}{S_{xx}}\right)}$$

With a p-value close to zero, in the ozone regression we end up rejecting the null hypothesis and conclude that each of our regression coefficients are "good." After calculating the upper and lower bounds of the confidence interval, another way to assess the importance of the regression coefficients is to determine whether or not the value 0 is contained within it. For example, if 0 is included in the confidence limits for the slope coefficient, it indicates you're dealing with a flat regression line whose slope is not different than 0.

In R, the confint() function provides the confidence intervals for a model fit with lm():

```
confint(lm.ozone.reg)
```

```
##                   2.5 %        97.5 %
## (Intercept) -183.222241  -110.768741
## Temp           1.966871     2.890536
```

Different confidence levels can be specified by adding the `level` = statement. For example, if we wanted to view the 80% confidence intervals instead of the default 95%:

```
confint(lm.ozone.reg, level = 0.80)
```

```
##                     10 %        90 %
## (Intercept) -170.568057 -123.422925
## Temp            2.128191    2.729215
```

8.5.4 Making predictions with a regression model

The ability to make predictions is one of the main reasons analysts perform regression techniques. For something that's relatively easy to measure (the independent variable), we can use regression to predict a hard-to-measure attribute (our dependent variable). After fitting a regression equation, you can write a customized function in R that applies your regression equation to a different data set or new observations that are added to an existing data set.

For a built-in approach in R, the `predict()` function allows you to apply regression equations to new data. For example, consider we have three new measurements of `Temp` and we would like to predict the corresponding level of `Ozone` based on the simple linear regression model we developed earlier. The three measurements are contained in the **ozone_new** data:

```
ozone_new <- tribble(
  ~Temp,
  65,
  75,
  82
)
```

The first argument of `predict()` is the model object that created your regression and the second argument is the data in which you want to apply the prediction. We'll use the `mutate()` function to add a column called `Ozone_new` that provides the predicted ozone value:

```
ozone_new %>%
  mutate(Ozone_pred = predict(lm.ozone.reg, ozone_new))
```

```
## # A tibble: 3 x 2
##     Temp Ozone_pred
##    <dbl>      <dbl>
## 1     65       10.9
## 2     75       35.2
## 3     82       52.2
```

A word of caution on the application of regression equations: they should only be used to make predictions about the population from which the sample was drawn. **Extrapolation** happens when a regression equation is applied to data from outside the range of the data used to develop it. For example, we would be extrapolating with our regression equation if we applied it to estimate ozone levels on days cooler than 50 degrees Fahrenheit. There are no observations in the data set used to develop the original regression lm.ozone.reg that were cooler than 50 degrees Fahrenheit.

As you seek to use other equations to inform your work in natural resources, it is important to understand the methods used in other studies that produced regressions. Having knowledge of the study area, population of interest, and sample size will allow you to determine which existing regression equations you can use and what the limitations are for applying equations that you develop to new scenarios.

8.5.5   Exercises

8.9 Print out the ANOVA table you developed earlier in question 8.6 that estimated the density of coarse woody debris based on the density of trees. Using manual calculations in R, calculate the R^2 value based on the sums of squares values provided in the ANOVA table.

8.10 Using manual calculations in R, calculate the R^2_{adj} value for the regression that estimated the density of coarse woody debris based on the density of trees.

8.11 Perform a literature search to find a study that performed linear regression using data from an organism that interests you. What were the independent and dependent variables used and what were the R_2 values that were reported? Are these findings surprising to you?

8.12 With the **air** data set, explore the relationship between ozone levels (Ozone) and wind speed (Wind).

   a.   Using ggplot(), make a scatter plot of these variables and add a linear regression line to the plot. Are ozone levels higher on windy or calm days?

b. Fit a simple linear regression model of Ozone (response variable) with Wind as the explanatory variable. Type the resulting coefficients for the $\hat{\beta}_0$ and $\hat{\beta}_1$.

c. Based on the summary output from the regression, calculate the Pearson correlation coefficient between Ozone and Wind.

d. Find the 90% confidence limits for the $\hat{\beta}_0$ and $\hat{\beta}_1$ regression coefficients.

e. Use your developed model to make ozone predictions for two new observations: when wind speed is 6 and 12 miles per hour.

8.13 A store wishes to know the impact of various pricing policies for milk on monthly sales. Using a data set named **milk**, they performed a simple linear regression between milk price (price) in dollars per gallon and sales in hundreds of gallons (sales). The following output results:

```
##
## Call:
## lm(formula = price ~ sales, data = milk)
##
## Residuals:
##      Min       1Q   Median       3Q      Max
## -0.50536 -0.15737 -0.02143  0.17388  0.42411
##
## Coefficients:
##             Estimate Std. Error t value Pr(>|t|)
## (Intercept)  5.65357    0.78876   7.168 9.54e-05 ***
## sales       -0.07054    0.02389  -2.953   0.0183 *
## ---
## Signif. codes:  0 '***' 0.001 '**' 0.01 '*' 0.05 '.' 0.1 ' ' 1
##
## Residual standard error: 0.2882 on 8 degrees of freedom
## Multiple R-squared:  0.5215, Adjusted R-squared:  0.4617
## F-statistic:  8.72 on 1 and 8 DF,  p-value: 0.01834
```

Using the R output, answer the following questions as TRUE or FALSE.

a. The price of milk was the independent variable in this simple linear regression.

b. There were 8 observations in the **milk** data set.

c. The Pearson correlation coefficient between milk price and number of sales was 0.7221.

d. For every dollar increase in the price of milk, sales decrease by seven cents.

e. For the median residual, the regression line predicts a value greater than its observed value (i.e., the regression "overpredicts" this value).

8.6 Model diagnostics in regression

Often in statistics we perform some analyses, then conduct some different analyses to see if it was even appropriate to conduct those analyses in the first place. Only then can we conclude whether or not it was appropriate to employ specific statistical methods. This is exactly what we'll do in regression, by visually inspecting our model's results and comparing them to what is expected.

Now that we've learned how to apply it in practice, we'll take a step back and visit the assumptions of linear regression.

8.6.1 Assumptions in regression

There are four core assumptions in simple linear regression. Assume we have n observations of an explanatory variable x and a response variable y. Our goal is to study or predict the behavior of y for given values of x. The assumptions are:

1. The means of the population under study are **linearly related** through the simple linear regression equation:

$$\mu_i = \beta_0 + \beta_1 x_i \text{ for } i = 1, 2, 3, ...n$$

In other words, you should see a linear pattern when you plot x and y on a scatter plot. If data do not appear linear, there are additional regression techniques that can be used to make inference with your data.

2. Each population has the same standard deviation (or **constant variance**). Written mathematically,

$$\sigma = \sigma_1 = \sigma_2 = \sigma_3 = ... = \sigma_n$$

The phenomenon of constant variance is also termed **homoscedasticity**. Hence, this assumption relies on the concept that values of the residuals do not increase or decrease as x increases. **Heteroscedasticity**, or non-constant variance, is problematic and is common with natural resources data.

3. The response measurements are independent normal random variables (**normality**). This assumption relies on the requirement that your response variable is normally distributed. As we saw in Chapter 2, the normal distribution is the foundation for statistical inference and its mean and standard deviation are denoted by μ and σ, respectively:

$$Y_i = N(\mu_i = \beta_0 + \beta_1 x_i, \sigma)$$

4. Observations (e.g., x_i and y_i) are **measured without error**. This may be an intuitive assumption, however, it can be overlooked in many applications. It is essential to have a thorough understanding of how the data were collected and processed before performing regression (and other statistical techniques). Two common problems in natural resources that lead to a violation of this assumption occur when (1) technicians measure experimental units with faulty equipment and (2) analysts are uncertain of the units associated with variables when performing statistical analyses (e.g., feet versus meters versus decimeters).

Many government agencies and large organizations maintain quality assurance and quality control (QA/QC) processes for data collected in their discipline. Two examples are:

• The US Geological Survey: `https://www.usgs.gov/products/data-and-tools/data-management/manage-quality`

• The US Department of Agriculture - Forest Service, Forest Inventory and Analysis: `https://www.fia.fs.fed.us/library/fact-sheets/data-collections/QA.pdf`

It is a good idea to become acquainted with any QA/QC processes that are used in your organization or discipline prior to analyzing with data.

8.6.2 Diagnostic plots in regression

8.6.2.1 Residual plots

You can check the conditions for regression inference by looking at graphs of the residuals, termed **residual plots**. Residual plots display the residual values $y_i - \hat{y}_i$ on the y-axis against the fitted values $\hat{y}_i$ on the x-axis. The ideal plot displays an even dispersion of residuals as you increase the fitted values $\hat{y}_i$. The residual should be evenly dispersed around zero, with approximately half of them positive and half of them negative. A trend line drawn through the residuals should be relatively flat and centered around 0.

In R, we will calculate predictions of ozone with the `predict()` function, and then manually calculate the residual value:

```
air <- air %>%
  mutate(Ozone_pred = predict(lm.ozone.reg, air),
         resid = Ozone - Ozone_pred)
```

We'll plot the residuals using `ggplot()` to assess our regression. Adding the default setting for `stat_smooth(se = F)` helps to visualize the trend in the residuals:

```
p.air.resid <- ggplot(air, aes(x = Ozone_pred, y = resid)) +
  geom_point() +
  ylab("Residual") +
  xlab("Fitted value") +
  stat_smooth(se = F)
p.air.resid
```

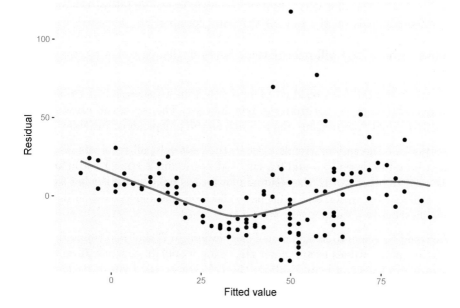

FIGURE 8.13 Residual plot for ozone in New York City.

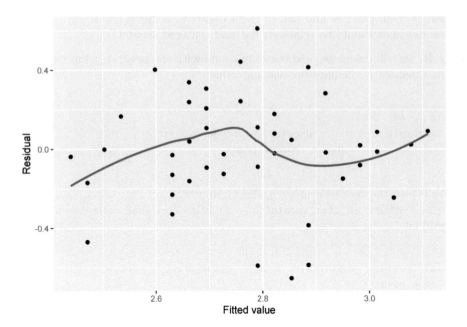

FIGURE 8.14 Residual plot for sepal width on Iris versicolor flowers.

Notice a few concerns with the ozone residual plot. First, there appears to be a trend in the plot with negative residuals in the middle of the data and positive residuals on the tails of the data, forming the U-shaped smoothed trend line. There also appear to be greater values for residuals as the fitted values increase. We will remedy these issues in the regression later on in the chapter.

Compare the previous residual plot to one from a regression fit to 50 observations of *Iris versicolor* from the **iris** data set. The regression estimates the width of sepal flowers using their length as the independent variable:

In this case, no serious trends exist in the residuals, all residuals are evenly distributed between -0.5 and 0.5 cm, and no residuals appear set apart from the rest. In other words, the residual plot for iris sepal width is what we strive for in linear regression and helps us to check off that our regression meets the qualifications for Assumption #2.

We could continue to make manual calculations in R that determine the residuals and other values useful in our regression model diagnostics. However, the **broom** package can help. Built on the **tidyverse**, the **broom** package summarizes key information about statistical objects and stores them in tibbles:

```
# install.packages("broom")
library(broom)
```

The augment() function from the package extracts important values after performing a regression. We can apply it to the lm.ozone.reg output and view the results in a tibble called air.resid:

```
air.resid <- augment(lm.ozone.reg)
air.resid
```

```
## # A tibble: 116 x 9
##    .rownames Ozone Temp .fitted .resid  .hat .sigma  .cooksd .std.resid
##    <chr>     <int> <int>  <dbl>  <dbl>  <dbl>  <dbl>    <dbl>      <dbl>
## 1  1            41    67   15.7   25.3  0.0200  23.7 0.0119        1.08
## 2  2            36    72   27.9   8.13  0.0120  23.8 0.000719      0.345
## 3  3            12    74   32.7  -20.7  0.0101  23.7 0.00393      -0.879
## 4  4            18    62   3.58   14.4  0.0330  23.8 0.00651       0.618
## 5  6            28    66   13.3   14.7  0.0222  23.8 0.00447       0.627
## 6  7            23    65   10.9   12.1  0.0246  23.8 0.00339       0.518
## 7  8            19    59  -3.70   22.7  0.0430  23.7 0.0215        0.979
## 8  9             8    61   1.16   6.84  0.0361  23.8 0.00162       0.294
## 9  11            7    74   32.7  -25.7  0.0101  23.7 0.00605      -1.09
## 10 12           16    69   20.6  -4.59  0.0162  23.8 0.000313     -0.195
## # ... with 106 more rows
```

The tibble contains a number of columns:

- .rownames: the identification number of the observation,
- Ozone and Temp: the values for the dependent (y) and independent (x) variable, respectively,
- .fitted: the predicted value of the response variable $\hat{y}_i$,
- .resid: the residual, or $y_i - \hat{y}_i$,
- .hat: the diagonals of the hat matrix from the regression, indicating the influence of observations,
- .sigma: the estimated residual standard deviation when the observation is dropped from the regression,
- .cooksd: Cook's distance, a value indicating the influence of observations, and
- .std.resid: the standardized residuals

Below we'll discuss more of these values in depth and plot them to evaluate trends.

8.6.2.2 Standardized residuals

In addition to the residual plot, you can also check the conditions for regression by looking at their standardized values. **Standardized residuals** represent the residual value $y_i - \hat{y}_i$ and divide them by their standard deviation. Whereas the residual values are in the units of your response variable, standardized residuals correct for differences by understanding the variability of residuals.

Think back to the normal distribution and the empirical rule. If the data are drawn from a normal distribution, we know that approximately 95% of observations will fall between two standard deviations of the mean. For standardized residuals, we should expect most values should be within ± 2. We can add reference lines to visualize this with the ozone data:

We can see five observations with a standardized residual greater than 2, which may indicate some adjustments needed in our regression model.

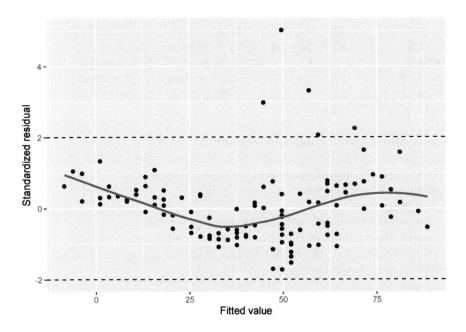

FIGURE 8.15 Standardized residuals for ozone in New York City.

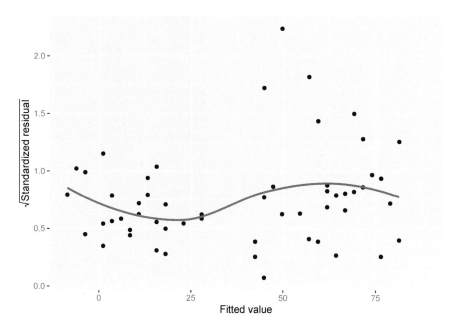

FIGURE 8.16 Square root of standardized residuals for ozone in New York City.

8.6.2.3 Square root of standardized residuals

By taking the **square root of the standardized residuals**, you rescale them so that they have a mean of zero and a variance of one. This eliminates the sign on the residual, with large residuals (both positive and negative) plotted at the top and small residuals plotted at the bottom.

In implementation, the trend line for this plot should be relatively flat. In the ozone example, there is some nonlinearity to the trend line. In particular, note the approximately five observations that continue to stand out as we visualize different representations of the residuals in this regression.

8.6.2.4 Cook's distance

Cook's distance is a measurement that allows us to examine specific observations and their influence on the regression. Specifically, it quantifies the influence of each observation on the regression coefficients. On the graph, the observation number is plotted along the x-axis with the distance measurement plotted along the y-axis.

In implementation, the Cook's values indicate influential data points and can be used to determine where more samples might be needed with your data.

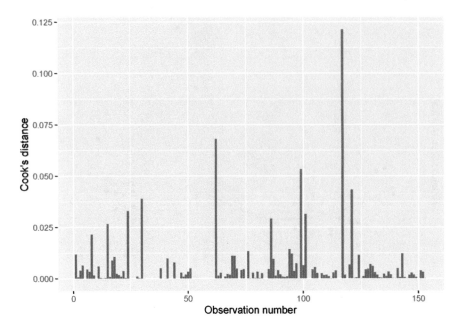

FIGURE 8.17 Cook's distance for ozone in New York City.

Plotting the values as a bar graph allows you to easily spot influential data points.

For example, observation numbers 117 and 62 present the highest Cook's values (0.121 and 0.067, respectively). It is worth checking these influential data points to ensure they are not data entry errors. Open the **air** data set and you'll see: while these two observations do not appear to be errors in data entry, they are the two largest recorded measurements of ozone (168 and 135 ppb, respectively). What likely makes these specific points influential is that they occurred on days with relatively mild temperatures (81 and 84 degrees Fahrenheit, respectively).

8.6.2.5 The diagPlot() function

The series of diagnostics plots shown above for the ozone regression can be replicated with the diagPlot() function below. The function requires a regression object termed model and uses ggplot() to stylize them:

```
diagPlot <- function(model){
# Residuals
p.resid <- ggplot(model, aes(x = .fitted, y = .resid)) +
```

```
  geom_point() +
  geom_hline(yintercept = 0, col = "black", linetype = "dashed") +
  ylab("Residual") +
  xlab("Fitted value") +
  stat_smooth(se = F)

# Standardized residuals
p.stdresid <- ggplot(model, aes(x = .fitted, y = .std.resid)) +
  geom_point() +
  geom_hline(yintercept = 2, col = "black", linetype = "dashed") +
  geom_hline(yintercept = -2, col = "black", linetype = "dashed") +
  ylab("Standardized residual") +
  xlab("Fitted value") +
  stat_smooth(se = F)

# Square root of standardized residuals
p.srstdresid <- ggplot(model, aes(x = .fitted,
                                   y = sqrt(abs(.std.resid)))) +
  geom_point() +
  ylab(expression(sqrt("Standardized residual"))) +
  xlab("Fitted value") +
  stat_smooth(se = F)

# Cook's distance
p.cooks <- ggplot(model, aes(x = seq_along(.cooksd), y = .cooksd)) +
  geom_bar(stat = "identity") +
  ylab("Cook's distance") +
  xlab("Observation number")

  return(list(p.resid = p.resid, p.stdresid = p.stdresid,
              p.srstdresid = p.srstdresid, p.cooks = p.cooks))
}
```

We can apply the function to the data set created by the `augment()` function in **broom**, i.e., the data with the residual values and other columns:

```
p.diag <- diagPlot(air.resid)
```

Extract one of the plots by connecting the function name and plot type with $. Or, add the diagnostic plots in a 2x2 panel by using the **patchwork** package:

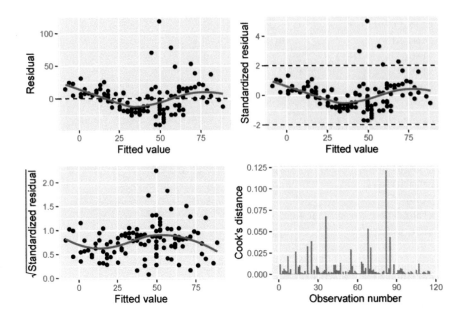

FIGURE 8.18 Panel graph showing four diagnostic plots of the ozone regression model.

```
install.packages("patchwork")
library(patchwork)
```

```
(p.diag$p.resid | p.diag$p.stdresid) /
(p.diag$p.srstdresid | p.diag$p.cooks)
```

8.6.3 How to remedy poor regressions: transformations

After inspecting the diagnostic plots for your regression, you may notice some issues with the performance of your regression model. Fortunately, many techniques exist that can remedy these situations so that you result in a statistical model that meets assumptions. Some of these techniques involve different adjustments within linear regression, or more advanced regression techniques can be evaluated.

The residual plot reveals a lot about the performance of a linear regression model. The appearance of the residuals can be thought of in the context of the assumptions of linear regression presented earlier in this chapter. For example, consider a residual plot with the following appearance:

```
## # A tibble: 81 x 5
##     ID       my_x resid1 my_rand    resid
##     <chr>   <dbl>  <dbl>   <dbl>    <dbl>
##  1 ID_1    -10      110    0.195   1.49
##  2 ID_1     -9.75   105.   0.937  78.2
##  3 ID_1     -9.5     99.8  0.292   9.17
##  4 ID_1     -9.25    94.8  0.990  73.8
##  5 ID_1     -9       90    0.343  10.9
##  6 ID_1     -8.75    85.3  0.233  -0.0853
##  7 ID_1     -8.5     80.8  0.662  33.4
##  8 ID_1     -8.25    76.3  0.150  -8.59
##  9 ID_1     -8       72    0.375   6.98
## 10 ID_1     -7.75    67.8  0.877  39.5
## # ... with 71 more rows
```

A U-shaped trend exists in the residuals, indicating some concerns with the regression. Likewise, our residual plot of the ozone-temperature regression

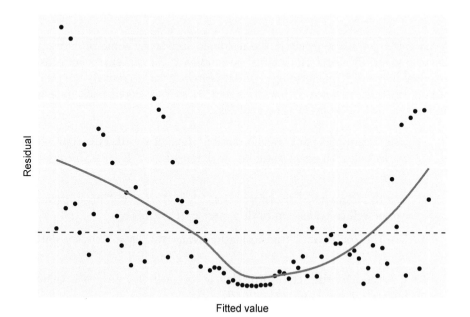

FIGURE 8.19 Simulated residual plot showing violation of the linear relationship in regression.

displays a similar trend. This patterns violates *Assumption 1*: there is not a linear relationship between the two variables and a nonlinear trend exists. To remedy regressions that appear like this, consider using nonlinear or polynomial regression techniques (discussed in Chapter 9).

Next, consider a residual plot with the following appearance. While the trend line for the residuals may be centered around zero, the magnitude of the residuals increases as the fitted values increase:

```
## # A tibble: 81 x 5
##     ID      my_x resid1 my_rand   resid
##     <chr> <dbl>  <dbl>    <dbl>   <dbl>
##  1 ID_1   0       0       7.70    0
##  2 ID_1   0.125   0.141  -0.987  -0.139
##  3 ID_1   0.25    0.312  -2.88   -0.900
##  4 ID_1   0.375   0.516  12.1     6.23
##  5 ID_1   0.5     0.75    9.94    7.45
##  6 ID_1   0.625   1.02  -13.4   -13.6
##  7 ID_1   0.75    1.31   -3.86   -5.06
##  8 ID_1   0.875   1.64    3.11    5.10
##  9 ID_1   1       2       1.48    2.97
## 10 ID_1   1.12    2.39   -1.35   -3.22
## # ... with 71 more rows
```

A "megaphone" shape exists in the residuals, indicating some problems with the regression. This pattern violates *Assumption 2*: the variance is not constant as values increase. Transforming your data can be a way to remedy the problem of non-constant variance. A few "tips and tricks" are noted here, depending on the types of variables in your analysis:

1. Use the **square-root** transformation when all $y_i > 0$. This approach works well with count data, i.e., with integers $i = 1, 2, 3, ..., n$: $y_i^2* = \sqrt{(y_i)}$.

2. Use the **logarithm** transformation when all $y_i > 0$. This approach works well to transform both y_i and x_i: $y_i^* = log(y_i)$.

3. Use the **reciprocal** transformation when you have non-zero data: $y_i^* = 1/y_i$.

4. Use the **arc sine** transformation on proportions, i.e., with values between 0 and 1: $y_i^* = arcsin\sqrt{y_i}$.

Note that after transforming variables, your interpretation of the results will change because the units of your variable have been altered. This is another

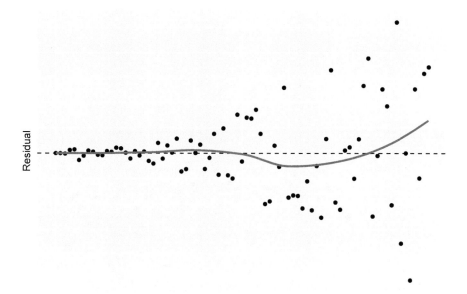

FIGURE 8.20 Simulated residual plot showing violation of the constant variance in regression.

reminder to be mindful of units and as a best practice, always type them into the labels on your axes for any figures and tables you produce.

Finally, consider a residual plot that looks generally good with randomness around 0, except for one or two "outlier" observations:

When you observe a few residuals that stand out from the rest, a violation of normality exists. Another graph that can examine the normality of your response variable is the **quantile-quantile plot**, or Q-Q plot. The Q-Q plot displays the kth observation against the expected value of the kth observation. For any mean and standard deviation, you would expect a straight line to result from data which are normally distributed. In other words, if the data on a normal Q-Q plot follow a straight line, we are confident that we don't have a serious violation of the assumption that the model deviations are normally distributed.

In R, the Q-Q plot can be generated using stat_qq() and stat_qq_line(). As an example, we can plot the ozone measures from the **air** data set:

```
ggplot(air, aes(sample = Ozone)) +
  stat_qq() +
  stat_qq_line()
```

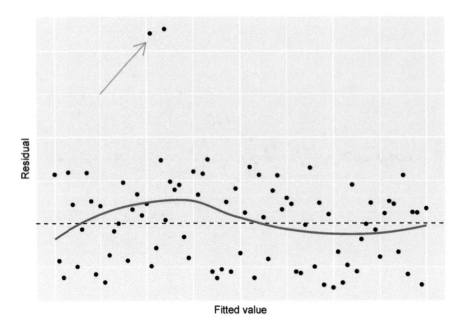

FIGURE 8.21 Simulated residual plot showing violation of normality in regression.

In the diagnostic plots for the regression on ozone measurements, we have noticed some issues that are also apparent in the Q-Q plot. The overall trend line of the values in the Q-Q plot show "sweeps" that do not follow a straight line. To remedy this violation of normality, check data entry and consult other regression techniques depending on the nature of your data.

8.6.4 Exercises

8.15. Recall the regression you developed earlier in question 8.6 that estimated the density of coarse woody debris based on the density of trees.

    a. With the regression object, run the `augment()` function from the **broom** package to obtain the data set with residual values and other variables. Write R code to calculate the mean and standard deviations of all residual values.

    b. Inspect the regression output by plotting the diagnostic plots using the `diagPlot()` function. Investigate the plots of residuals, standardized residuals, square root of standardized residuals, and Cook's distance. Write two to three sentences that describe the diagnostic

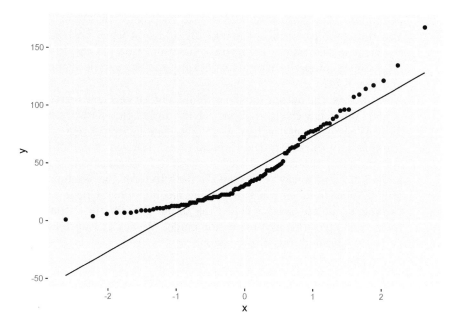

FIGURE 8.22 A quantile-quantile plot of ozone measurements from New York City.

plots and any concerns you might have about the fit of the regression.

c. Create a quantile-quantile plot for the density of coarse woody debris and describe the pattern you see. After your assessment of the plot, are you confident in stating that the data are distributed normally?

8.16 Recall back to question 8.12 which fits the regression of New York City ozone levels with wind speed.

a. Provide the output for the residual plot from this regression model. What is the general trend in the residual plot? Which core assumption is violated in this regression and what indicates this violation?

b. One way to remedy the violation observed here is to perform a **polynomial regression**. Polynomial regression is a type of linear regression in which the relationship between the independent variable x and the dependent variable y is modeled as an nth degree polynomial. In other words, we can add a squared term of our independent variable Wind to create a second-degree polynomial model. Create a new variable in the **air** data set using the mutate() function that squares the value of Wind. Then, perform a multiple linear regression using both Wind and Windsq to predict the ozone level. You can combine them in the code by adding a Wind + Windsq to the right of the tilde ~ in the lm() function. How do the R^2 and R^2_{adj} values compare with this model and the one you developed in question 8.12?

c. Now, provide the output for the residual plot for this new regression model. What has changed about the characteristics of the residual plot? What can you conclude about the assumptions of regression?

d. Another way to remedy the residual plots observed in 8.16.a is to log-transform the response variable Ozone. Log-transforming data is often used as a way to change the data to meet the assumptions of regression (and get a better-looking residual plot). Create a new variable in the **air** data set called log_Ozone. Provide the output for the residual plot this new regression model. What can you conclude about the assumptions of regression?

8.7 Summary

Regression is one of the most commonly used tools in the field of statistics. To estimate one attribute from another, it is first useful to evaluate the correlation between variables. If a correlation exists, simple linear regression techniques can be used to estimate one variable from another. The analysis of variance table can help us to understand the proportion of the model that is explained by the regression (i.e., the regression sums of squares) versus the error (i.e., the residual sums of squares).

Just as statistical inference allows us to assess differences between means and proportions, inference can be performed with regression coefficients. These inferential procedures include performing hypothesis tests on the intercept and slope and providing confidence intervals. Understanding the stability of the regression coefficients will allow us to make reliable predictions for the response variable. Combining our inference with a visual inspection of regression output through diagnostic plots, we can determine whether or not our regression model is appropriate and meets each of the core four assumptions.

The lm() function in R provides a large volume of output for linear regression models and will be used extensively in future chapters. Combining the numerical output with visual assessments of residuals and other model diagnostics, we can evaluate the performance of our regression model. With a thorough grasp of regression, we can begin learning more about more advanced estimation techniques to understand phenomenon in the natural world.

8.8 Reference

Christensen, D.L., B.R. Herwig, D.E. Schindler, S.R. Carpenter. 1996. Impacts of lakeshore residential development on coarse woody debris in north temperate lakes. *Ecological Applications* 6: 1143–1149.

9

Multiple regression

9.1 Introduction

We have considered in detail the linear regression model in which a mean response is related to a single explanatory variable. **Multiple linear regression** will allow us to use two or more independent variables and employ least squares techniques to make appropriate inference. We've learned how to conduct hypothesis tests and calculate confidence intervals for coefficients such as β_1 and β_2 in a simple linear regression framework (i.e., the intercept and slope). This chapter will provide us the tools to understand the slopes associated with additional independent variables, e.g., β_2, β_3, and β_4.

If you have more information that might help you to make a better prediction, why not use it? This is the value of multiple regression and why it is widely used in natural resources. There are several benefits of multiple linear regression, including:

- it allows us to use *several variables* to explain the variation in a single dependent variable,
- we can isolate the effect of a *specific independent variable* and see how it changes the predictions of a dependent variable, and
- we can *control for other variables* to see relationships between all variables.

Fortunately, the concepts of least squares apply as in simple linear regression . We can use many of the same functions in R for multiple linear regression that we're familiar with.

9.2 Multiple linear regression

There are many problems in which knowledge of more than one explanatory variable is necessary to obtain a better understanding and more accurate prediction of a particular response. We will continue to use the concepts of least squares to minimize the residual sums of squares (*SS(Res)*) in multiple linear regression. The general model is written as:

$$y = \beta_0 + \beta_1 x_1 + \beta_2 x_2 + \ldots + \beta_p x_p + \varepsilon$$

where β_0 is the intercept, β_p are the "slopes" for each independent variable, and x_p are the pth independent variables. The least squares estimates for the coefficients $\hat{\beta}_p$ involve more computations compared to simple linear regression. For $p = 2$ independent variables, we can calculate the coefficients $\hat{\beta}_p$ if we know:

- the Pearson correlation coefficient r between all variables y, x_1, and x_2,
- the mean values of all variables $\bar{y}, \bar{x}_1$, and $\bar{x}_2$, and
- the sample standard deviations of all variables s_y, s_{x_1}, and s_{x_2}.

The formulas to calculate the least squares estimates of a multiple linear regression with $p = 2$ dependent variables is

$$\hat{\beta}_0 = \bar{y} - \hat{\beta}_1 \bar{x}_1 - \hat{\beta}_2 \bar{x}_2$$

$$\hat{\beta}_1 = \left(\frac{r_{y,x_1} - r_{y,x_2} r_{x_1,x_2}}{1 - (r_{x_1,x_2})^2} \right) \left(\frac{s_y}{s_{x_1}} \right)$$

$$\hat{\beta}_2 = \left(\frac{r_{y,x_2} - r_{y,x_1} r_{x_1,x_2}}{1 - (r_{x_1,x_2})^2} \right) \left(\frac{s_y}{s_{x_2}} \right)$$

To understand the principles of multiple linear regression, consider data collected on peregrine falcons (*Falco peregrinus*). Data were simulated based on information presented in Kéry (2010) and include the following variables:

- `wingspan`: the wingspan length measured in cm,
- `weight`: weight, measured in g,
- `tail`: tail length, measured in cm, and
- `sex`: sex of the falcon (female or male).

These attributes were measured on 20 falcons, presented in the **falcon** data set from the **stats4nr** package:

```
## # A tibble: 6 x 4
##   wingspan weight  tail sex
##      <dbl>  <dbl> <dbl> <chr>
## 1       96    950  15.5 F
## 2       97   1100  16.5 F
```

FIGURE 9.1 Peregrine falcon. Image: Christopher Watson.

```
## 3      109    1295  17.6 F
## 4      110    1350  18.2 F
## 5       97    1150  16.6 F
## 6      102    1380  17.5 F
```

Falcons are a sexually dimorphic species, indicating that females are larger than males. The following plot shows the distribution of falcon weights by sex using geom_density():

```
ggplot(falcon, aes(x = weight, col = sex)) +
        geom_density() +
        labs(x = "Falcon weight (g)")
```

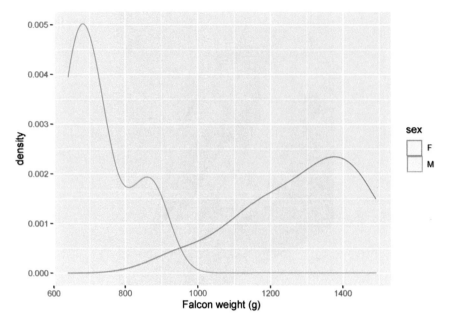

With multiple linear regression you likely have lots of variables, and it's a good practice to start with visualizing them. The **GGally** package in R allows you to produce a matrix of scatter plots, distributions, and correlations among multiple variables. This is very useful in multiple regression when we might be interested in looking at a large number of variables and their relationships. The ggpairs() function provides this, and we can see the differences between male (M) and female (F) falcons:

```
install.packages("GGally")
library(GGally)



ggpairs(falcon, aes(col = sex))
```

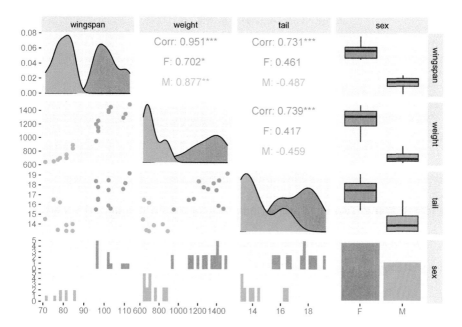

We are interested in predicting the wingspan length of these falcons. In addition to looking at the `ggpairs()` output, we find the correlation between `wingspan` and `weight` to be 0.95:

```
cor(falcon$wingspan, falcon$weight)
```

```
## [1] 0.9512555
```

So, there is a strong positive correlation between falcon wingspan length and weight. However there is still some variability that is not accounted for when investigating these two variables (Fig. 9.2).

Consider another independent variable that might be useful: the length of the falcon's tail. In addition to determining the correlation between wingspan and tail length (0.73), we also know the correlation between the two independent variables tail length and weight (0.74), and the correlation between wingspan and the "interaction" between tail length and weight (0.94):

```
cor(falcon$wingspan, falcon$tail)
```

```
## [1] 0.7313432
```

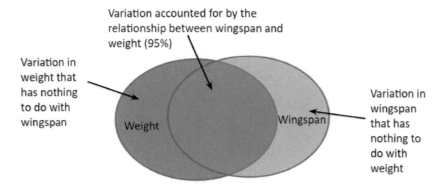

FIGURE 9.2 Two-way interactions between falcon weight and wingspan.

```
cor(falcon$weight, falcon$tail)
```

```
## [1] 0.7385734
```

```
cor(falcon$wingspan, falcon$tail*falcon$weight)
```

```
## [1] 0.9406831
```

These three-way interactions are summarized in Fig. 9.3, and you can see that adding additional independent variable captures more of the variability associated with the response variable.

As in simple linear regression, the method of least-squares for multiple linear regression chooses $\hat{\beta}_0, \hat{\beta}_1, ...\hat{\beta}_p$ to minimize the sum of the squared deviations $(y_i - \hat{y}_i)^2$. The regression coefficients reflect the unique association of each independent variable with the response variable y and are analogous to the slopes in simple linear regression.

9.2.1 Implementation in R

In R, multiple linear regression can be specified by adding a + between all independent variables. We can perform a multiple linear regression using both weight and tail length to predict the wingspan of falcons:

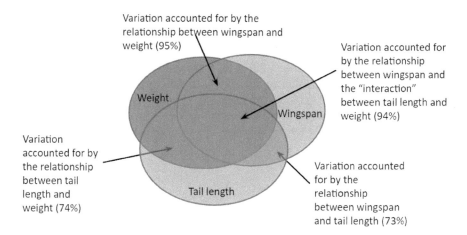

FIGURE 9.3 Three-way interactions between falcon weight, wingspan, and tail length.

```
falcon.lm <- lm(wingspan ~ weight + tail, data = falcon)
summary(falcon.lm)
```

```
##
## Call:
## lm(formula = wingspan ~ weight + tail, data = falcon)
##
## Residuals:
##     Min      1Q  Median      3Q     Max
## -5.3103 -3.0005 -0.4342  2.3996  7.2011
##
## Coefficients:
##              Estimate Std. Error t value Pr(>|t|)
## (Intercept) 48.514650   8.642648   5.613 3.10e-05 ***
## weight       0.035743   0.004342   8.232 2.47e-07 ***
## tail         0.408272   0.708697   0.576    0.572
## ---
## Signif. codes:  0 '***' 0.001 '**' 0.01 '*' 0.05 '.' 0.1 ' ' 1
##
## Residual standard error: 3.945 on 17 degrees of freedom
## Multiple R-squared:  0.9067, Adjusted R-squared:  0.8957
## F-statistic: 82.61 on 2 and 17 DF,  p-value: 1.753e-09
```

We can see that regression coefficients indicate falcon wingspans are longer on heavier birds with longer tail lengths. Note the *p*-value of 0.572 indicates tail length as a non-significant variable in the model, a finding we'll revisit shortly.

Two other functions from the **broom** package allow you to extract important values from regression output. The `tidy()` function extracts the key values associated with the regression coefficients and places them in an R data frame stored as a tibble:

```
library(broom)
tidy(falcon.lm)
```

```
## # A tibble: 3 x 5
##    term         estimate std.error statistic    p.value
##    <chr>           <dbl>     <dbl>     <dbl>      <dbl>
## 1 (Intercept)     48.5      8.64       5.61  0.0000310
## 2 weight         0.0357   0.00434      8.23  0.000000247
## 3 tail            0.408    0.709      0.576  0.572
```

The `glance()` function provides an R data frame containing the summary statistics of the model, such as R^2 and R^2_{adj}

```
glance(falcon.lm)
```

```
## # A tibble: 1 x 12
##    r.squared adj.r.squared sigma statistic    p.value    df
##        <dbl>         <dbl> <dbl>     <dbl>      <dbl> <dbl>
## 1      0.907         0.896  3.94      82.6    1.75e-9     2
## # ... with 6 more variables: logLik <dbl>, AIC <dbl>,
## #   BIC <dbl>, deviance <dbl>, df.residual <int>,
## #   nobs <int>
```

Similar to what we did in simple linear regression, the `augment()` function can be applied to the multiple regression object to view the residuals:

```
falcon.resid <- augment(falcon.lm)
head(falcon.resid)
```

```
## # A tibble: 6 x 9
##    wingspan weight tail .fitted .resid  .hat .sigma .cooksd .std.resid
##       <dbl>  <dbl> <dbl>   <dbl>  <dbl> <dbl>  <dbl>   <dbl>      <dbl>
## 1       96    950  15.5    88.8   7.20 0.0574   3.62  0.0718       1.88
```

```
## 2      97   1100  16.5    94.6   2.43  0.0522   4.02 0.00736      0.633
## 3     109   1295  17.6    102.   7.01  0.0856   3.63 0.108        1.86
## 4     110   1350  18.2    104.   5.80  0.115    3.76 0.106        1.56
## 5      97   1150  16.6    96.4  0.603 0.0544   4.06 0.000474      0.157
## 6     102   1380  17.5    105.  -2.99  0.105    3.99 0.0251      -0.800
```

We can run the `diagPlot()` function presented in Chapter 8 to visualize the model's diagnostic plots:

```
library(patchwork)

p.diag <- diagPlot(falcon.resid)

(p.diag$p.resid | p.diag$p.stdresid) /
(p.diag$p.srstdresid | p.diag$p.cooks)
```

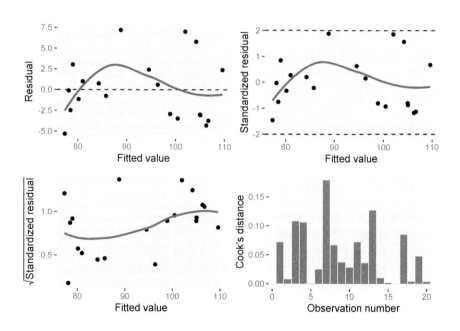

The residual and other diagnostic plots look satisfactory for this multiple linear regression. There is a slight trend in larger values for residuals as fitted values increase, which may provide some justification for transforming the data. For the purpose of illustration, we'll continue to investigate the output from the `falcon.lm` regression.

TABLE 9.1 The analysis of variance (ANOVA) table in multiple linear regression.

Source	Degrees of freedom	Sums of squares	Mean square	F
Regression	p	SS(Reg)	MS(Reg) = SS(Reg)/p	MS(Reg) / MS(Res)
Residual	n-p-1	SS(Res)	MS(Res) = SS(Res)/n-p-1	-
Total	n-1	TSS	-	-

9.2.2 Inference in multiple regression

Recall we can calculate confidence intervals to inform us about the population means of the regression coefficients. To understand the confidence intervals of $\beta_0, \beta_1, ...\beta_p$, we rely again on the t distribution, with $n-p-1$ degrees of freedom. The confint() function also provides the confidence intervals for multiple linear regression objects:

```
confint(falcon.lm)
```

```
##                  2.5 %     97.5 %
## (Intercept) 30.280257 66.7490436
## weight       0.026582  0.0449045
## tail        -1.086948  1.9034918
```

We can use the ANOVA table to partition the variance of a multiple linear regression model. The greatest difference compared to simple linear regression is the degrees of freedom for the regression and residual components, which depends on p, the number of predictors used in the regression.

The anova() function in R produces the ANOVA table. Consider the falcon wingspan regression:

```
anova(falcon.lm)
```

```
## Analysis of Variance Table
##
## Response: wingspan
##
##            Df  Sum Sq Mean Sq  F value    Pr(>F)
## weight      1 2566.08 2566.08 164.8921 3.546e-10 ***
```

```
## tail        1    5.16    5.16   0.3319    0.5721
## Residuals 17  264.56   15.56
## ---
## Signif. codes:   0 '***' 0.001 '**' 0.01 '*' 0.05 '.' 0.1 ' ' 1
```

Note that the function provides two rows for `weight` and `tail` that can be added together to calculate the $SS(Reg)$ and $MS(Reg)$ values.

In multiple regression, the F-statistic follows an $F(p, n - p - 1)$ distribution and tests the null hypothesis

$$H_0 : \beta_1 = \beta_2 = ... = \beta_p$$

versus the alternative hypothesis

$$\text{At least one } \beta_p \neq 0$$

A rejection of the null hypothesis and a significant p-value in multiple linear regression should be interpreted carefully. This does not mean that **all p explanatory variables** have a significant influence on your response variable, only that **at least one variable** does. The falcon output is a good example of this: the overall F-statistic is 82.61 indicating a rejection of the null hypothesis for the multiple linear regression. However, we see that `weight` is the only significant variable in the model at the $\alpha = 0.05$ level and `tail` is not.

When developing regression models, you should subscribe to the Albert Einstein quote: "Everything should be made as simple as possible, but no simpler." Statisticians and data analysts term these **parsimonious models**, or models that achieve maximum prediction capability with the minimum number of independent variables needed. The next section discusses how to perform linear regression when you have a large number of independent variables available.

9.2.3   Exercises

9.1 The **fish** data set contains measurements on fish caught in Lake Laengelmavesi, Finland (Puranen 1917). The data include measurements from seven species where the weight, length, height, and width of fish were measured. The variables are defined as:

- `Species`: (Fish species name)
- `Weight`: (Weight of the fish [grams])
- `Length1`: (Length from the nose to the beginning of the tail [cm])
- `Length2`: (Length from the nose to the notch of the tail [cm])
- `Length3`: (Length from the nose to the end of the tail [cm])
- `Height`: (Maximal height as % of Length3)

- Width: (Maximal width as % of Length3)

   a. Read in the data set from the **stats4nr** package and explore the
      data using the ggpairs() function from the **GGally** package. Add a
      different color to your plot for each of the seven species. What do you
      notice about the relationships between the three length variables?
      Which fish species has the largest median value for width?

   b. Create a new data set called **perch** that contains only observations
      of perch species using the filter() function. Then, use lm() to
      create a simple linear regression with this data set to predict Weight
      using Length3 as an independent variable (the length from the nose
      to the end of the tail). Use summary() on the model to investigate
      its performance. How much would a perch with a Length3 of 30 cm
      weigh?

   c. Apply the diagPlot() function to the simple linear regression model.
      What is the general trend in the residual plot? Which assumption
      of regression is violated here and what indicates this violation?

   d. To remedy the regression, create a new variable in the **perch** data
      set called Length3sq that squares the value of Length3. In addi-
      tion, create a new variable by log-transforming the response vari-
      able Weight. Then, perform a multiple linear regression using both
      Length3 and Length3sq to predict the log-transformed weight. Then,
      provide the output for the residual plot for this model. What has
      changed about the characteristics of the residual plot? What can
      you conclude about the assumptions of this regression?

   e. Use the glance() function from the **broom** package to compare the
      adjusted R-squared values between the simple and multiple linear
      regression models.

9.2 This question performs a multiple linear regression using the **falcon** data
set.

   a. "Dummy" variables, or indicator variables, are dichotomous indica-
      tors of a variable used in a regression. Create a new dummy variable
      in the **falcon** data set that creates a numeric 1,0 variable that repre-
      sents whether a falcon is male or female. Then, perform a multiple
      linear regression predicting wingspan using weight and the dummy
      variable for the sex of the falcon. Do you reject or fail to reject the
      null hypothesis that all slopes β_p are equal to zero at the $\alpha = 0.05$
      level?

   b. Use the tidy() function on the multiple linear regression to extract
      the regression coefficients.

c. Provide 85% confidence intervals for the regression coefficients in this model.

9.3   All subsets regression

In many natural resources data sets, dozens of independent variables might be available that are related to your response variable of interest. We might be interested in predicting our response variables by testing the performance of linear regressions using all of these variables. However, there may be a lot of different combinations of all possible regressions to do, so many that we don't want to fit an lm() model for each of them. That would take too much time!

The **leaps** package in R searches for the best subsets of all variables that might be used in a multiple linear regression:

```
install.packages("leaps")
library(leaps)
```

We can use the regsubsets() function from the package to run all subsets regression predicting a response variable of interest. For example, consider the **falcon** data set. We create the falcon.leaps object using the function, which appears similar to the lm() function:

```
falcon.leaps <- regsubsets(wingspan ~ weight + tail + factor(sex),
                           data = falcon)
```

We can plot the R_{adj}^2 values for all subsets using the plot() function:

```
plot(falcon.leaps, scale="adjr2")
```

In the plot, each row represents a model. The models are ranked from the highest R_{adj}^2 on the top to the lowest on the bottom. In this falcon example, the regression subset indicates that a model with an intercept, weight, and sex of the falcon performs best (e.g., it provides the highest R_{adj}^2). The shading color is related to the significance level of the variables, with darker colors indicating lower p-values.

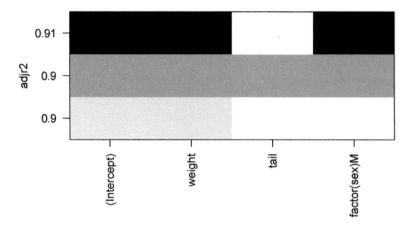

FIGURE 9.4 Plot showing adjusted R-squared values from the leaps package.

9.3.1 Stepwise regression

Stepwise regression techniques involve using an automated process to select which variables to include in a model. Two stepwise techniques involve forward (beginning with a null model and adding variables) and backward regression (beginning with all variables in a model and subsequently removing them). The general approach in a stepwise regression is to set a significance level α for deciding when to enter a predictor into the model. We also set a significance level for deciding when to remove a predictor from the model. The values chosen for α are typically 0.15 or less.

To select a subset of predictors from a larger set, we'll use the `step()` function in R to help with this. First, we'll fit a null model for the **falcon** data set using no predictors for `wingspan`:

```
null <- lm(wingspan ~ 1, data = falcon)
```

Then, we'll fit a full model, using all possible predictors of `wingspan`:

```
full <- lm(wingspan ~ weight + tail + factor(sex), data = falcon)
```

The step() function can be used to implement forward regression using direction = "forward". In this approach, we begin with no variables in the model
and proceed forward, adding one variable at a time. The R output displays
the forward regression process:

```
step(null, scope = list(lower = null, upper = full),
     direction = "forward")
```

```
## Start:  AIC=101.09
## wingspan ~ 1
##
##               Df Sum of Sq     RSS     AIC
## + weight       1    2566.1  269.72  56.033
## + factor(sex)  1    2332.0  503.79  68.529
## + tail         1    1516.8 1319.04  87.778
## <none>                     2835.80 101.087
##
## Step:  AIC=56.03
## wingspan ~ weight
##
##               Df Sum of Sq    RSS    AIC
## + factor(sex)  1    36.974 232.75 55.084
## <none>                     269.72 56.033
## + tail         1     5.165 264.56 57.646
##
## Step:  AIC=55.08
## wingspan ~ weight + factor(sex)
##
##        Df Sum of Sq    RSS    AIC
## <none>              232.75 55.084
## + tail  1  0.036654 232.71 57.081


##
## Call:
## lm(formula = wingspan ~ weight + factor(sex), data = falcon)
##
## Coefficients:
##  (Intercept)       weight   factor(sex)M
##      65.6764       0.0282        -6.4069
```

At each step, output shows the **Akaike's information criterion (AIC)**,
a measure of prediction error that measures the quality of a model. Models
with lower AIC values are ideal, and a model will be preferred over another
if the difference in AIC is less than two units. Also shown are the sums of

squares and residual sums of squares as each variable is added to the model. The coefficients listed at the end of the output indicate that in addition to the intercept, weight and sex are the variables chosen in the forward regression.

The step() function can also be used to implement backward regression (direction = "backward"). In this approach, we start with all potential variables in the model and proceed backward, removing one variable at a time:

```
step(full, direction = "backward")
```

```
## Start:  AIC=57.08
## wingspan ~ weight + tail + factor(sex)
##
##                 Df Sum of Sq    RSS    AIC
## - tail           1     0.037 232.75 55.084
## <none>                        232.71 57.081
## - factor(sex)    1    31.846 264.56 57.646
## - weight         1   261.916 494.63 70.161
##
## Step:  AIC=55.08
## wingspan ~ weight + factor(sex)
##
##                 Df Sum of Sq    RSS    AIC
## <none>                        232.75 55.084
## - factor(sex)    1    36.974 269.72 56.033
## - weight         1   271.044 503.79 68.529


##
## Call:
## lm(formula = wingspan ~ weight + factor(sex), data = falcon)
##
## Coefficients:
##  (Intercept)       weight   factor(sex)M
##      65.6764       0.0282        -6.4069
```

Note that the backward regression approach identifies the same variables as the forward regression.

> **DATA ANALYSIS TIP**: Generally, if you have a very large set of predictors (say, greater than five) and are looking to identify a few key variables in your model, forward regression is best. If you have a modest number of predictors (less than five) and want to eliminate a few, use backward regression.

9.3.2 Evaluating multiple linear regression models

Multiple linear regression models can become unwieldy as you add numerous independent variables to your model. For example, you may be using too many variables in your model such that they are correlated with one another. This is termed **multicollinearity**, i.e., when one variable is held constant, another is in effect also held constant. Multicollinearity is a topic that needs to be addressed in analyses of multiple linear regression.

The consequences of not addressing multicollinearity can be serious. These include (1) a final model with regression coefficients that have large uncertainty and (2) an inaccurate representation of the sums of squares for regression and residual components. The primary solutions for avoiding multicollinearity include (1) removing some of the independent variables and (2) fitting separate models with each independent variable.

The **variance inflation factor (VIF)** is a useful metric that evaluates the degree of multicollinearity in a regression model. The greater the VIF, the greater the collinearity between variables. A VIF for a single independent variable is obtained using the R^2 of the regression of that variable compared to all other independent variables. Generally, VIF values greater than 10 indicate a need to reevaluate which independent variables are used in the regression.

The **car** package was developed to accompany a textbook on regression and contains useful functions for regression (Fox and Weisberg 2009). We will install the package and load it with `library()` so that we can look at some of the functions that are contained within it.

```
install.packages("car")
```

```
library(car)
```

We can determine VIF using the `vif()` function available in the car package for the full model with three variables predicting falcon wingspan:

```
vif(full)
```

```
##      weight        tail factor(sex)
##    5.492267    2.495839   6.045090
```

The VIF values are all less 10, indicating no serious multicollinearity present.
You can see the values for VIF decrease when you evaluate the reduced model
using only weight and sex:

```
reduced <- lm(wingspan ~ weight + factor(sex), data = falcon)
```

```
vif(reduced)
```

```
##      weight factor(sex)
##    5.328972    5.328972
```

9.3.3 Exercises

9.3 Recall the **elm** data set on cedar elm trees located in Austin, Texas to
answer the following questions.

  a. Read in the data set and explore the data using the ggpairs()
     function. What do you notice about the distribution of DIA and HT?
     What is the correlation coefficient between HT and CROWN_DIAM_WIDE?

  b. Create a new dummy variable in the **elm** data set that creates
     a {1,0} variable that represents how much light a tree receives. If
     it is an open-grown or dominant tree (i.e., a CROWN_CLASS_CD of 1
     or 2), the tree will have a 1 for the dummy variable. These trees
     are receiving a lot of light. If it is a co-dominant, intermediate, or
     overtopped tree (i.e., a CROWN_CLASS_CD of 3, 4, or 5), the tree
     will have a 0 for the dummy variable. These trees have some or no
     light available to them. How many trees are in each category?

  c. We might be interested in predicting elm tree heights (HT) using
     all of these variables. Use the regsubsets() function to run all
     subsets regression predicting HT using the independent variables
     DIA, CROWN_HEIGHT, CROWN_DIAM_WIDE, UNCOMP_CROWN_RATIO, and the
     dummy variable you made for CROWN_CLASS_CD (a factor vari-
     able). Visualize the results by plotting the R^2_{adj} for all models using

the `plot()` function. What is the R^2_{adj} for this model? Which variables are chosen in the best performing model?

d. Fit both forward and backward regressions considering all variables to predict HT. Which variables are selected by the forward and backward regression techniques? Are these the same sets of variables? Explain why they may or may not be the same.

e. Calculate the variance inflation factor for the variables identified in the backward selection process. Are you concerned that multicollinearity is present in the model?

9.4 Let's revisit the **perch** data set. Fit a regression model that predicts the weight of perch using the independent variables Length1, Length2, and Length3 and calculate the variance inflation factor. Are you concerned with multicollinearity and/or variance inflation in this model?

9.4 Summary

Multiple regression techniques allow us to consider more than one independent variable to make a prediction on our response variable of interest. Similar to simple linear regression, the `lm()` function can be used to fit multiple regression models. Stepwise regression techniques allow us to identify which variables are best to include in our regression models through an automated process, without the need to fit subsets of models using different combinations of independent variables. This allows us to "prune" our variables to result in only the most important ones, however, some cautions need to be recognized when fitting multiple linear models.

Multicollinearity is a widespread concern in many biological and natural resources data sets. Any analysis using multiple predictor variables in a modeling framework should evaluate multicollinearity. For making comparisons between multiple linear regression models, evaluation statistics such as R^2_{adj} and AIC are useful metrics.

9.5 References

Fox, J., S. Weisberg. 2019. *An R companion to applied regression, 3rd ed.* Sage. 608 p.

Kéry, M. 2010. *Introduction to WinBUGS for ecologists.* Academic Press. 302 p.

Puranen, J. 1917. Fish catch data set. *Journal of Statistics Education, Data Archive*: Available at: `http://jse.amstat.org/datasets/fishcatch.txt`

10

Analysis of variance

10.1 Introduction

There are lots of practical situations where there are more than two populations of interest that we want to evaluate. If we had just two populations, we could run a two-sample t-test to compare their means. **Analysis of variance**, abbreviated ANOVA or sometimes AOV, extends these methods to problems involving more than two populations.

The ANOVA process compares the variation from specific sources with the variation among individuals that should be similar. In particular, ANOVA tests whether several populations have the same means by comparing how far apart they are with how much variation there is within a sample.

This chapter will discuss how to use the ANOVA model to make inference between population means and draw conclusions about their differences. Similar to concepts in regression, we'll construct an ANOVA table and examine residual plots to evaluate the assumptions about ANOVA models.

10.2 One-way analysis of variance

R.A. Fisher introduced the term **variance** and proposed its formal analysis in a 1918 article (Fisher 1928). His first application of ANOVA occurred a few years later and the method became widely known after being included in his book *Statistical Methods for Research Workers* (Fisher 1925), which was updated in numerous subsequent editions of the book. Much of the early work with ANOVA used agricultural trials and experiments in England, and the method was adopted in medicine in the mid-1900s. (Unfortunately, Fisher's work was closely integrated into the science of eugenics, a difficult point in the history of the discipline of statistics.)

The purpose of ANOVA is to assess whether the observed differences among sample means are statistically significant. Consider an example data set that

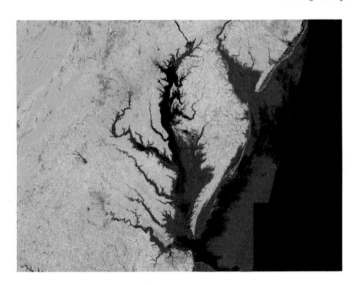

FIGURE 10.1 Map of Chesapeake Bay, eastern United States. Image: Landsat/NASA.

investigates iron levels in Chesapeake Bay, a large estuary in the eastern United States.

Iron levels were measured at several water depths in Chesapeake Bay (Sananman and Lear 1961). The researchers were interested to know if changing water depths influence the amount of iron content in the water.

Experimenters took three measurements at six water depths: 0, 10, 30, 40, 50, and 100 feet. The response variable was iron content, measured in mg/L. The data are contained in **iron** from the **stats4nr** package:

```
library(stats4nr)

iron
```

```
## # A tibble: 18 x 2
##     depth  iron
##     <dbl> <dbl>
## 1      0 0.045
## 2      0 0.043
## 3      0 0.04
## 4     10 0.045
## 5     10 0.031
## 6     10 0.043
```

```
##  7      30 0.044
##  8      30 0.044
##  9      30 0.048
## 10      40 0.098
## 11      40 0.074
## 12      40 0.154
## 13      50 0.117
## 14      50 0.089
## 15      50 0.104
## 16     100 0.192
## 17     100 0.213
## 18     100 0.224
```

We can make a box plot of the iron level at various water depths:

```
library(tidyverse)
```

```
p.iron <- ggplot(iron, aes(factor(depth), iron)) +
  geom_boxplot()+
  ylab("Iron content (mg/L)") +
  xlab("Water depth (feet)")
p.iron
```

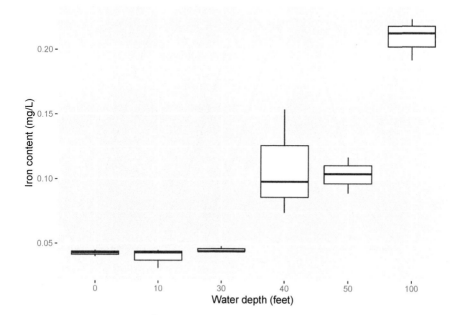

So, it appears that changing water depths influences iron content in the bay. But what else can we say about these relationships? That is, is there a statistical difference in the iron content at different water depths? Alas, the ANOVA method will allow us to examine the influence of factors (e.g., water depths) on a response variable of interest (e.g., iron levels).

First, understanding a few key definitions is essential to understanding the notation associated with ANOVA.

- **Treatments** or **levels** are the individual experimental conditions defining a population, e.g., the water depths.
- **Experimental units** are the individual objects influenced by a treatment, e.g., the water samples.
- **Responses** are measurements obtained from the experimental units, e.g., iron content.

The parameters of the ANOVA model are the population means $\mu_1, \mu_2, ..., \mu_i$ with the common standard deviation σ across all i treatments. We can use the sample mean to estimate each μ_i. Similar to the two-sample t-test, we can pool the standard deviations across all of our treatments or levels. The one-way ANOVA model analyzes data where chance variations are normally distributed.

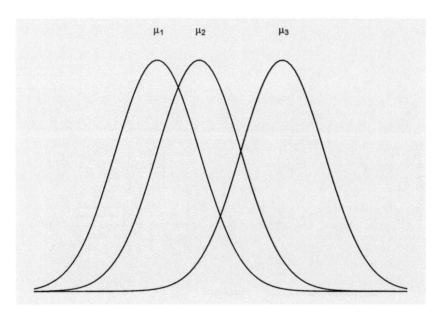

FIGURE 10.2 Three different distributions that can be examined in an analysis of variance.

The null hypothesis in the ANOVA is that there are no differences among the means of the populations. The alternative hypothesis is that there is some difference across population means. That is, not all means are equal:

$$H_0 : \mu_1 = \mu_2 = ... = \mu_i$$

$$H_A : \text{at least one mean different from the rest}$$

The basic conditions for inference in ANOVA are that we have a random sample from each population and that each population is normally distributed. We will perform the ANOVA F-test to examine the hypothesis.

Like all inference procedures, ANOVA is valid only in some circumstances. The conditions under which we can use ANOVA include:

- We have i independent simple random samples, one from each population. We measure the same response variable for each sample.
- The ith population has a normal distribution with unknown mean μ_i.
- All the populations have the same standard deviation σ, whose value is unknown.

10.2.1 Partitioning the variability in ANOVA

Just like in regression, recall that the sums of squares represent different sources of variation in the data. In ANOVA, these values include the total (*TSS*), between group (*SSB*), and within group sums of squares (*SSW*). Large values of *SSB* reflect large differences in treatment means, while small values of *SSB* likely indicate no differences in treatment means. The calculations are:

$$TSS = \sum (x_{ij} - \bar{x})^2$$

$$SSB = \sum n_i (\bar{x}_i - \bar{x})^2$$

$$SSW = \sum (n_i - 1) s_i^2$$

where x_{ij} is the jth observation from the ith treatment, $\bar{x}$ is the overall sample mean, n_i is the number of observations in treatment i, $\bar{x}_i$ is the sample mean of the ith treatment, and s_i is the sample standard deviation of treatment i.

We can use the ANOVA table to make inference about the factors of interest under study. In the table, p is the number of treatments or independent comparison groups minus 1.

TABLE 10.1 The one-way analysis of variance (ANOVA) table.

Source	Degrees of freedom	Sums of squares	Mean square	F
Within	p	SSW	MSW = SSW/p	MSB / MSW
Between	n-p-1	SSB	MSB = SSB/n-p-1	-
Total	n-1	TSS	-	-

So, the ANOVA F test can be determined with the test statistic $F_0 = \frac{MSB}{MSW}$. Similar to a test comparing variances of two populations, the ANOVA F-test is always an upper-tailed test.

In the **iron** data set we can perform a one-way ANOVA with the lm() function and view the ANOVA table with the anova() function. First, we'll convert the water depth variable (currently stored as a number) to a factor variable. This is because the water depths are labeled as numbers, but they represent categorical variables in our treatment of them in the ANOVA:

```
iron <- iron %>%
  mutate(depth.fact = as.factor(depth))

iron.aov <- lm(iron ~ depth.fact, data = iron)

anova(iron.aov)
```

```
## Analysis of Variance Table
##
## Response: iron
##             Df   Sum Sq   Mean Sq  F value    Pr(>F)
## depth.fact   5 0.064802 0.0129605   35.107 9.248e-07 ***
## Residuals   12 0.004430 0.0003692
## ---
## Signif. codes:  0 '***' 0.001 '**' 0.01 '*' 0.05 '.' 0.1 ' ' 1
```

In R, variation between groups (*SSB*) is labeled by the name of the grouping factor, i.e., depth.fact. Variation within groups (*SSW*) is labeled *Residuals*. In the case of the iron data, we see that with a small p-value of the ANOVA F-test (9.248e-07), we reject the null hypothesis that all means are equal and conclude that at least one mean differs from the rest. Hence, iron content is not equal across all water depths.

We can run the `diagPlot()` function presented in Chapter 8 to visualize the diagnostic plots from ANOVA output, along with supporting functions from the **broom** and **patchwork** packages:

```
library(broom)
library(patchwork)

iron.resid <- augment(iron.aov)

p.diag <- diagPlot(iron.resid)

(p.diag$p.resid | p.diag$p.stdresid) /
(p.diag$p.srstdresid | p.diag$p.cooks)
```

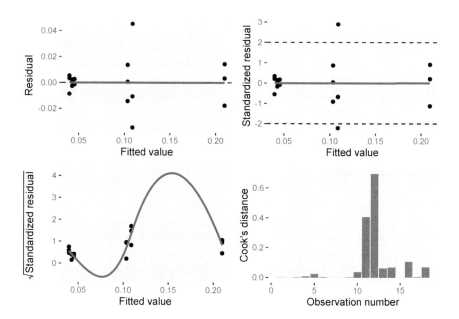

Interpreting the diagnostic plots is a little tricky because our iron data are "grouped" at different water depths. However, the diagnostic plots look satisfactory for this ANOVA model.

10.2.2 Exercises

10.1 Cannon et al. (1997) measured the number of tree species after logging in several rainforests in Borneo. Treatments included forests that were never logged, logged one year ago, and logged eight years ago.

   a. Read in the **logging** data set from the **stats4nr** package and find out how many observations are in the data.

   b. Create a violin plot showing the distribution of the number of tree species across the three treatments. Describe in a few sentences the differences you observe across treatments.

   c. Perform a one-way analysis of variance that determines the effects of logging treatment on tree species diversity using a level of significance $\alpha = 0.05$. What is the conclusion of your analysis?

   d. Run a series of diagnostic plots for this analysis using the `diagPlot()` function. After inspecting the plots, is there any aspect of the results that gives you pause or urges you to transform the data?

10.2 Analysis of variance is widely used in natural resources. Describe a study you might be aware of that uses ANOVA, or do a literature search to find a peer-reviewed article that uses ANOVA in your discipline. Try searching "analysis of variance" and "your discipline/study area." Include the following in your response:

- What did the study examine (e.g., the response variable and units) and what were the treatments or levels used?
- What were the significant (or non-significant) results of the study?
- Provide a citation for your chosen article.

10.3   Two-way analysis of variance

The purpose of a **two-way analysis of variance** is to understand if two independent variables and their interaction have an effect on the dependent variable of interest. Treatment groups are formed by making all possible combinations of the two factors. For example, if the first factor has three levels and the second factor has two levels, there will be $3 \times 2 = 6$ different treatment groups.

For example, consider an experiment that plants three different varieties of winter barley at four different dates during the year. We have 12 treatment groups and measure the yield after a growing season. Say we take $n = 10$ samples from each treatment group. In a two-way ANOVA the degrees of freedom are calculated as follows:

- There are $4 - 1 = 3$ degrees of freedom for the planting date.
- There are $3 - 1 = 2$ degrees of freedom for the barley variety.

- There are $2 * 3 = 6$ degrees of freedom for the interaction between planting date and variety.

An important attribute of a two-way ANOVA is that there are three null hypotheses under investigation. In the case of the barley-planting date experiment, these hypotheses are:

- The population means of the first factor are equal (planting date).
- The population means of the second factor are equal (variety).
- There is no interaction between the two factors (planting date*variety).

Due to the three hypotheses in a two-way ANOVA, this indicates there are three F-tests to conduct for each hypothesis. For the two **main effects**, you can investigate each independent variable one at a time. This is an approach similar to a one-way analysis of variance.

Two-way ANOVAs also include an evaluation of the **interaction effect**, or the effect that one factor has on the other factor. For example, some barley varieties might grow faster if planted at different dates, and this interaction is important to separate out from their individual effects. Two-way ANOVAs also incorporate an error effect, calculated by determining the sum of squares within each treatment group.

The key assumptions in a two-way ANOVA analysis include the following:

- The populations from which the samples were obtained must be normally or approximately normally distributed.
- The samples must be independent.
- The variances of the populations must be equal.
- The groups must have the same sample size.

The two-way ANOVA table has several more rows to take into account the two main effects and their interaction. Like all ANOVA tables we've looked at, the mean square values are taken by dividing the sums of squares by their degrees of freedom. The important consideration is the number of levels or treatments in main effect A (a) and B (b).

As an example, consider the **moths** data set from the **stats4nr** package. The data contain the number of spruce moths found in trees after 48 hours (Brase and Brase 2017). Moth traps were placed in four different locations in trees (top, middle, lower branches, and ground). Three different types of lures were placed in the traps (scent, sugar, chemical). There were five observations for each treatment at each level:

TABLE 10.2 The two-way analysis of variance (ANOVA) table.

Source	Degrees of freedom	Sums of squares	Mean square	F
Main effect A	a-1	SS(A)	MS(A) = SS(A)/a-1	MS(A)/MSW
Main effect B	b-1	SS(B)	MS(B) = SS(B)/b-1	MS(B)/MSW
Interaction effect	(a-1)(b-1)	SS(AB)	MS(AB) = SS(AB)/(a-1)(b-1)	MS(AB)/MSW
Error (Within group)	n-ab	SSW	MSW = SSW/(n-ab)	-
Total	n-1	TSS	-	-

```
library(stats4nr)
moths
```

```
## # A tibble: 60 x 3
##     Location Lure   Moths
##     <chr>    <chr>  <dbl>
##  1 Top      Scent     28
##  2 Top      Scent     19
##  3 Top      Scent     32
##  4 Top      Scent     15
##  5 Top      Scent     13
##  6 Top      Sugar     35
##  7 Top      Sugar     22
##  8 Top      Sugar     33
##  9 Top      Sugar     21
## 10 Top      Sugar     17
## # ... with 50 more rows
```

To visualize the data, we'll create a box plot showing the distribution of the number of moths by trap location, using fill = Lure within the aes() statement in ggplot(). This provides a different colored box plot for each lure treatment:

```
ggplot(moths, aes(Location, Moths, fill = Lure)) +
  geom_boxplot() +
```

FIGURE 10.3 Western spruce budworm moth. Image: Utah State University

```
ylab("Number of moths") +
xlab("Location")
```

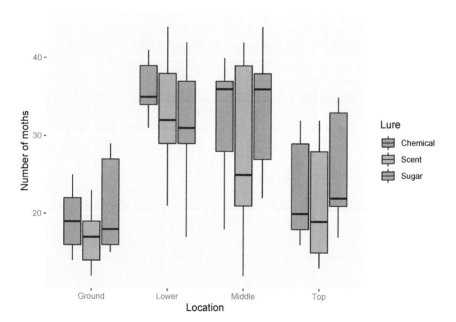

We can see that there are fewer spruce moths counted at ground level than other locations. The null hypotheses for this test are:

- The population means of Location are equal.
- The population means of Lure are equal.
- There is no interaction between Location and Lure.

To perform two-way ANOVAs in R, we can use the `lm()` function. We could type `Moths ~ Location + Lure + Location*Lure` for the model, however, R will recognize the shorthand `Moths ~ Location*Lure` to specify the two main effects plus their interaction:

```
moths.aov <- lm(Moths ~ Location * Lure, data = moths)
anova(moths.aov)
```

```
## Analysis of Variance Table
##
## Response: Moths
##                Df  Sum Sq  Mean Sq  F value    Pr(>F)
## Location        3 1981.38   660.46  10.4503 2.094e-05 ***
## Lure            2  113.03    56.52   0.8943    0.4156
## Location:Lure   6  114.97    19.16   0.3032    0.9322
## Residuals      48 3033.60    63.20
## ---
## Signif. codes:  0 '***' 0.001 '**' 0.01 '*' 0.05 '.' 0.1 ' ' 1
```

At the $\alpha = 0.05$ level of significance, the results of our two-way ANOVA hypotheses are:

- The null hypothesis that population means of Location are equal is rejected.
- The null hypothesis that population means of Lure are equal is accepted.
- The null hypothesis that there is no interaction between Location and Lure is accepted.

So, we can conclude that the placement of the trap within a tree influences the number of moths captured while the type of lure used does not. There is not an interaction between the location of the trap and the lure type.

10.3.1 Exercises

10.3 An experiment was conducted with four levels of main effect A and five levels of main effect B. Answer the following questions.

    a. What are the degrees of freedom for the F statistic that is used to test for the interaction in this analysis?

    b. Explain how hypothesis testing in a two-way ANOVA differs from that in a one-way ANOVA. Write two to three sentences.

10.4 The **CO2** data set is built into R and contains carbon dioxide uptake in plant grasses at two locations (Quebec and Mississippi) and two treatments (chilled and nonchilled). Load the data by typing`CO2 <- tibble(CO2)` and complete the following questions:

    a. Subset the **CO2** data set so that you are only working with a CO2 concentration of 500 mL/L. How many total observations are in the data?

    b. Write R code to perform a two-way ANOVA at a level of significance $\alpha = 0.10$ that determines the effect of location and treatment on carbon dioxide uptake rates (`uptake`). What are the null hypotheses of this analysis?

    c. What is the outcome of the ANOVA in terms of each of the hypothesis tests, i.e., are the null hypotheses accepted or rejected?

10.5 The **butterfly** data set is found in the **stats4nr** package and contains the wing length (in mm) of 60 mourning cloak butterflies from Kéry (2010). Butterfly data were simulated at different elevations and populations, with the hypothesis that butterflies from different populations that were hatched at different elevations would be subject to different climate conditions and predators. Load the data and complete the following questions:

    a. How many levels of elevation and different populations were evaluated in the data?

    b. Create a box plot that shows the distribution of wing lengths by the different elevations and populations. What do you notice about the distributions of wing lengths?

    c. Write R code to perform a two-way ANOVA at a level of significance $\alpha = 0.10$ that determines the effect of population and elevation on wing lengths. What are the null hypotheses of this analysis?

    d. What is the outcome of the ANOVA in terms of each of the hypothesis tests, i.e., are the null hypotheses accepted or rejected?

10.4   Multiple comparisons

In the example with iron levels, we concluded that at least one mean at a certain water depth was different from another. But doesn't it make sense that we'll want to know *which specific water depths* are significantly different? After performing an ANOVA and concluding that you reject the null hypothesis, i.e., that at least one population mean is different, you should be left wanting to know more information about these differences.

TABLE 10.3 Changes in experiment-wise Type I error with different number of comparisons (m).

m	Experiment-wise error rate
1	0.0500
2	0.0975
5	0.2262
10	0.4012
15	0.5367

The method of **multiple comparisons** allows us to assess significant differences across specific treatment groups, assuming we've already found a significant overall effect in the ANOVA F-test. While it might be intuitive to use what we learned in Chapter 4 and conduct pooled t-tests, in this approach, this would test $H_0 : \mu_i = \mu_j$ versus $H_A : \mu_i \neq \mu_j$ for each pair μ_i and μ_j with $i \neq j$. However, a problem arises when performing multiple tests. The level of significance α represents the probability of rejecting the null hypothesis when in fact it is true. To combat this, we will calculate an error-rate that adjusts the Type I error depending on how many levels or treatments we're comparing.

This **experiment-wise error rate**, which we denote α_E, reflects all comparisons under consideration in our ANOVA. Hence, α_E represents the probability of making at least one false rejection when comparing multiple populations. If you consider m different independent comparisons,

$$\alpha_E = 1 - (1 - \alpha)^m \geq \alpha$$

For $\alpha = 0.05$ and various levels of m, the experiment-wise Type 1 error probabilities change. Note that the experiment-wise Type I error increases as the number of independent comparisons increases.

So, statisticians quickly realized that while the ANOVA F-test was important to understand whether or not a treatment influences a response variable, also important is finding out which specific treatments differ from one another with their influence on the response. Multiple comparison procedures were developed to compensate for increases in α_E. Multiple comparison procedures, also known as pairwise tests of significance, should be carried out **if and only if** the null hypothesis in the ANOVA was rejected. In other words, if your ANOVA F-test did not yield a significant result, you can conclude your analysis. There is no need to look for significant differences among treatments.

If your ANOVA F-test indicated to reject the null hypothesis and you conclude that at least one mean is different from the rest, a multiple comparison analysis is appropriate. There are many specific methods for doing multiple comparisons that range from more to less conservative approaches. Commonly used multiple comparison tests used in natural resources fields include Tukey's honestly significant difference (HSD), Scheffé, and Bonferroni. A more conservative test is one that requires greater differences between two means to declare them significantly different. To determine which multiple comparison test is appropriate for you, consult the literature in your discipline and seek feedback from colleagues.

Regardless of the specific multiple comparison test chosen, each are variants on the two-sample t-test. Standard deviations are pooled across treatments being evaluated and are adjusted for the number of tests being evaluated simultaneously. For illustration, we'll use an example using the Bonferroni multiple comparison procedure.

10.4.1   One-way ANOVA

In R, the `pairwise.t.test()` function calculates all possible comparisons between different groups. It can also conduct multiple comparisons so that we can determine which locations are significantly different from others.

We'll use `pairwise.t.test()` to obtain the p-values which test for significance differences among the iron content across water depths. For a one-way ANOVA, we'll provide the function three arguments: the response variable (`iron`), the variable representing the treatment or level (`depth.fact`), and the p-value adjustment method for the multiple comparison test. In our case we'll use the Bonferroni method (`p.adj = "bonferroni"`):

```
pairwise.t.test(iron$iron, iron$depth.fact, p.adj = "bonferroni")


##
##  Pairwise comparisons using t tests with pooled SD
##
## data:  iron$iron and iron$depth.fact
##
##      0        10       30       40       50
## 10  1.00000  -        -        -        -
## 30  1.00000  1.00000  -        -        -
## 40  0.01825  0.01302  0.02472  -        -
## 50  0.03359  0.02380  0.04578  1.00000  -
## 100 2.7e-06  2.2e-06  3.2e-06  0.00048  0.00029
##
## P value adjustment method: bonferroni
```

The output provides a matrix of *p*-values that make multiple comparisons across the different depths where iron content was sampled. At the level of significance $\alpha = 0.05$, nearly all comparisons can be considered significantly different *except*:

- 0 and 10 feet,
- 0 and 30 feet,
- 10 and 30 feet, and
- 40 and 50 feet.

This makes intuitive sense because the box plot of data shown previously indicates clusters of iron levels between 0 and 30 feet, 40 and 50 feet, and 100 feet. As you will notice, the output from `pairwise.t.test()` is manageable with a small number of treatments being evaluated, however, it can quickly become messy to interpret the matrix table with an increasing number of treatments in your study. Fortunately, functions in the **agricolae** package can help to distill the output from multiple comparisons:

```
install.packages("agricolae")
library(agricolae)
```

The `LSD.test()` function from **agricolae** also provides output on the multiple comparison tests. We'll create the `lsd.iron` R object to store this output:

```
lsd.iron <- LSD.test(iron.aov, "depth.fact", p.adj = "bonferroni")
lsd.iron$groups
```

```
##              iron groups
## 100 0.20966667      a
## 40  0.10866667      b
## 50  0.10333333      b
## 30  0.04533333      c
## 0   0.04266667      c
## 10  0.03966667      c
```

The output here is much more detailed, containing summaries such as averages, confidence limits, and quantile values for iron content at each of the different water depths. Under $groups, you'll see the mean value for each water depth along with lowercase letters. Different letters denote significant differences across water level depths, while groups with the same letter indicate no

TABLE 10.4 Example table that summarizes iron content data, with different lowercase letters denoting significant differences across water depths.

Water depth (ft)	Mean iron content (mg/L)	SD iron content (mg/L)	Significance
0	0.0427	0.0025	a
10	0.0397	0.0076	a
30	0.0453	0.0023	a
40	0.1087	0.0411	b
50	0.1033	0.0140	b
100	0.2097	0.0163	c

significant differences. Again, the clustered nature of the data are apparent, with groups of 0 and 30 feet (group c), 40 and 50 feet (group b), and 100 feet (group a).

10.4.2 Visualizing differences across groups

After looking at the R output, you will soon learn which treatments or levels are significantly different from others. However, visualizing your results in an effective manner can help your audience interpret your analysis and its findings. With ANOVAs that employ multiple comparison procedures, these can take the form of tables or figures.

A well-designed table with appropriate units can help to convey your results in a compact form. The table should include the treatment name and columns indicating mean and standard deviation values, with letters denoting significant differences.

You might notice that the letters *a* and *c* were switched in the table compared to what was shown in the LSD.test output. The specific letters are arbitrary so long as the grouping of levels is consistent. In addition, it makes for easier reading to begin with the letter *a* in the first row of the table.

Figures are also appropriate to display ANOVA and multiple comparison results. With a study containing a limited number of treatments, bar plots work well to visualize results. To make a bar plot using ggplot(), we can first make a tidy data set that summarizes the raw data and provides the number of observations and mean and standard deviation values. This is accomplished by grouping the data by treatment (i.e., depth.fact()) and then summarizing:

```
iron_summ <- iron %>%
  group_by(depth.fact)   %>%
```

```
summarize(n.iron = n(),
          mean.iron = mean(iron),
          sd.iron = sd(iron))
```

```
iron_summ
```

```
## # A tibble: 6 x 4
##    depth.fact n.iron mean.iron sd.iron
##    <fct>       <int>     <dbl>   <dbl>
## 1 0               3    0.0427 0.00252
## 2 10              3    0.0397 0.00757
## 3 30              3    0.0453 0.00231
## 4 40              3    0.109  0.0411
## 5 50              3    0.103  0.0140
## 6 100             3    0.210  0.0163
```

A bar plot that shows mean values with error bars depicting variability shows the range of data across treatments. For the error bars, we can create an R object called `limits` that contains maximum and minimum values, or $\pm$ one standard deviation around the mean:

```
limits <- aes(ymax = mean.iron + sd.iron,
              ymin = mean.iron - sd.iron)
```

Our `ggplot()` code to make a bar plot will contain the following important elements:

- `geom_errorbar(limits, width = 0.25)` will add the error bars defined in `limits` that display $\pm$ one standard deviation around the mean. The `width()` statement controls the width of the error bar lines relative to the bars.
- `geom_text()` will add the letter denoting significant differences, which can be typed manually. Adding `vjust = -4` will vertically adjust the text labels so that their labels allow space between the error bars.
- `scale_y_continuous()` adjusts the scale of the y-axis, a trick that is often needed when adding error bars that extend the plotting area.

```
p.iron <- ggplot(iron_summ, aes(depth.fact, mean.iron)) +
  geom_bar(stat = "identity") +
  geom_errorbar(limits, width = 0.25) +
  ylab("Iron content (mg/L)") +
```

```
xlab("Water depth (feet)") +
geom_text(aes(label = c("a", "a", "a", "b", "b", "c")), vjust = -6) +
scale_y_continuous(limits = c(0, 0.3))
p.iron
```

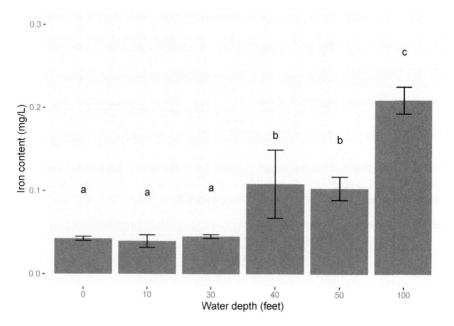

In this example, ± one standard deviation around the mean is displayed. Other commonly used error bars in natural resources include ± one or two standard errors around the mean. For studies with many treatments (e.g., > 10), faceting (i.e., creating a panel of graphs) or using color can help your reader visualize the results.

10.4.3 Two-way ANOVA

Referring back to the **moths** data set, you should have observed that trap location within the tree has an effect on the number of moths caught. Our ANOVA results indicated trap location had a significant effect on the number of spruce moths caught, but which specific locations differ? If we consider a level of significance of $\alpha = 0.025$, we can also use the pairwise.t.test() function to obtain the p-values which test for significance differences among the four locations.

```
pairwise.t.test(moths$Moths, moths$Location, p.adj = "bonferroni")
```

```
##
##   Pairwise comparisons using t tests with pooled SD
##
## data:  moths$Moths and moths$Location
##
##        Ground  Lower   Middle
## Lower  2.3e-05 -       -
## Middle 0.00044 1.00000 -
## Top    0.78828 0.00420 0.04791
##
## P value adjustment method: bonferroni
```

We can see that the middle and ground ($p = 0.00044$), lower and top ($p = 0.00420$), and lower and ground locations ($p = 2.3e - 05$) are significantly different from each other at the $\alpha = 0.025$ level.

We can also compare the result using the LSD.test() function:

```
lsd.moth <- LSD.test(moths.aov, "Location", p.adj = "bonferroni")
lsd.moth
```

```
## $statistics
##    MSerror Df      Mean       CV t.value      MSD
##       63.2 48  26.68333 29.79329 2.752023 7.988772
##
## $parameters
##          test   p.ajusted   name.t ntr alpha
##    Fisher-LSD  bonferroni Location   4  0.05
##
## $means
##             Moths      std  r      LCL      UCL Min Max  Q25 Q50  Q75
## Ground 19.06667 5.091543 15 14.93956 23.19378  12  29 15.5  18 22.5
## Lower  33.33333 7.499206 15 29.20622 37.46044  17  44 30.0  34 38.5
## Middle 31.00000 9.790666 15 26.87289 35.12711  12  44 23.5  36 38.5
## Top    23.33333 7.412987 15 19.20622 27.46044  13  35 17.5  21 30.5
##
## $comparison
## NULL
##
## $groups
```

TABLE 10.5 Example table that summarizes moth abundance data, with different lowercase letters denoting significant differences across water depths.

Trap location	Mean moths	SD moths	Significance
Ground	19.07	5.09	a
Lower	33.33	7.50	b
Middle	31.00	9.79	bc
Top	23.33	7.41	ac

```
##              Moths groups
## Lower   33.33333      a
## Middle  31.00000     ab
## Top     23.33333     bc
## Ground  19.06667      c
##
## attr(,"class")
## [1] "group"
```

The results for the **moths** data are more interesting than for the **iron** data. Whereas the iron content samples were clustered, the moth abundance at different traps located in the tree are more varied. Different letters denote similarities between traps located in the lower and middle portions of the tree (group a), middle and top (group b), and top and ground (group c).

We observe that traps placed in the lower portions of trees have the greatest average number of moths and the ground location the lowest. We can also summarize the moth abundance data to create a tidy data set to use in plotting. We will add a variable se.moths that calculates the standard error of the number of moths:

```
moths_summ <- moths %>%
  group_by(Location) %>%
  summarize(n.moths = n(),
            mean.moths = mean(Moths),
            sd.moths = sd(Moths)) %>%
    mutate(se.moths = sd.moths/sqrt(n.moths))

moths_summ
```

```
## # A tibble: 4 x 5
##    Location n.moths mean.moths sd.moths se.moths
##    <chr>      <int>      <dbl>    <dbl>    <dbl>
## 1 Ground        15       19.1     5.09     1.31
```

```
## 2 Lower          15       33.3       7.50       1.94
## 3 Middle         15       31         9.79       2.53
## 4 Top            15       23.3       7.41       1.91
```

We will create a bar plot with error bars that displays the mean number of spruce moths captured. Again, we'll create the minimum and maximum values for the error bars in `limits()`, but this time using $\pm$ one standard error of the mean. We also add letters denoting significant differences:

```
limits <- aes(ymax = mean.moths + se.moths,
              ymin = mean.moths - se.moths)

p.loc <- ggplot(moths_summ, aes(Location, mean.moths)) +
  geom_bar(stat = "identity") +
  geom_errorbar(limits, width = 0.25) +
  geom_text(aes(label = c("a", "b", "bc", "ac")), vjust = -4) +
  scale_y_continuous(limits = c(0, 50)) +
  ylab("Mean moth abundance") +
  xlab("Location of traps")
p.loc
```

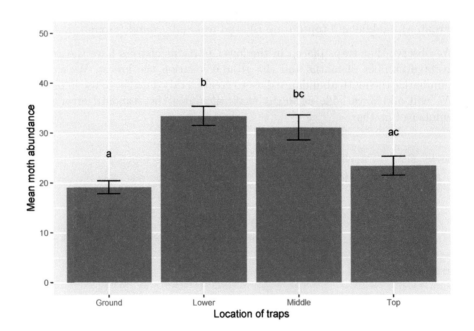

10.4.4 Exercises

10.6 Consider the one-way ANOVA conducted in question 10.1 with the **logging** data set using the LSD.test() to answer the following questions.

  a. Use pairwise.t.test() to obtain the p-values which test for significance differences across the three types of forests, using the Bonferroni multiple comparison test. Assuming a level of significance of $\alpha = 0.05$, which forests are significantly different from one other?
  b. Use the LSD.test() function from the **agricolae** package to determine which forests are significantly different from one another.
  c. Create a bar plot that shows the mean values of species richness in each of the forests with error bars denoting $\pm$ one standard deviation. Orient the bars horizontally by adding coord_flip() to your ggplot() code.

10.7 Use the LSD.test() function on the **butterfly** data to determine which populations show significantly different wing lengths. Conduct your test at a level of significance of $\alpha = 0.10$.

10.8 A number of multiple comparison tests are available to compare differences across treatments. Use pairwise.t.test() to obtain the p-values which test for significance differences among the four spruce moth trap locations using the *false discovery rate* multiple comparison test. This multiple comparison test examines the false discovery rate of treatment differences. (HINT: Type ?p.adjust to learn about other multiple comparison tests available in R.)

Assuming a level of significance $\alpha = 0.025$, which specific locations are significantly different from each other using the false discovery rate test? How do these compare to what we observed for the Bonferroni test earlier in the chapter? Would you consider the false discovery rate multiple comparison test to be more or less conservative compared to the Bonferroni test?

10.9 For the spruce moth analysis and your response to question 10.8, would you be interested in comparing differences across individual lure types, e.g., chemical vs scent vs sugar? Explain why or why not.

10.10 Make a new bar plot for the mean number of spruce moths by trap location that shows error bars for $\pm$ two standard errors of the mean of the number of spruce moths captured. Include the letters denoting significant differences. (HINT: You may need to adjust the scale on the y-axis and the location of the letters above the bars to make sure everything fits in your plot.)

10.5   Summary

The lm() function in R was familiar to us in simple linear regression to quantify the relationship between two quantitative variables. In ANOVA, our predictor variable represents a treatment or level and is categorical in nature. The ANOVA process determines if population means of the variable representing the treatment or level are different from one another.

The process of ANOVA does not reveal which specific means are different from one another, only that a difference exists. That is where multiple comparisons come into play. Multiple comparison procedures, which vary in terms of how much evidence is needed to observe a difference, examine which specific treatments or factors vary from one another. Showcasing ANOVA results using tables and figures helps to reveal the results of your analysis in an impactful way.

10.6   References

Brase, C.H., and C.P. Brase. 2017. *Understandable statistics: concepts and methods, 11th ed.* Cengage Learning, Boston, MA.

Cannon, C.H., Peart, D.R., Leighton, M. 1998. Tree species diversity in commercially logged Bornean rainforest. *Science* 281: 1366--1367.

Fisher, R.A. 1918. The correlation between relatives on the supposition of Mendelian inheritance. *Transactions of the Royal Society of Edinburgh* 52(2): 399--433.

Fisher, R.A. 1925. *Statistical methods for research workers.* Oliver and Boyd: Edinburgh.

Kéry, M. 2010. *Introduction to WinBUGS for ecologists.* Academic Press. 302 p.

Sananman, M., D. Lear. 1961. Iron in Chesapeake Bay waters. *Chesapeake Science* 2(3/4), 207–209.

11

Analysis of covariance

11.1 Introduction

Analysis of variance tests whether several populations have the same means by comparing how far apart they are and how much variation there is within a sample. Linear regression allows us to examine relationships between at least two quantitative variables. **Analysis of covariance** (ANCOVA) is a general linear model which blends concepts from ANOVA and regression. The ANCOVA technique evaluates whether the means of a dependent variable are equal across levels of a categorical independent variable, often a treatment that is controlled in an experiment or some other factor variable. The ANOVA process controls for the effects of other continuous variables that are not of primary interest, known as covariates.

This chapter will discuss how ANCOVA combines principles from ANOVA and regression, the assumptions of an ANCOVA model and how they can be assessed, and how to apply ANCOVA techniques to natural resources data sets.

11.2 Analysis of covariance

You can think about ANCOVA as an application when we add a continuous covariate in an ANOVA analysis. These covariates are not generally a part of the manipulation of an experiment, but they may have an influence on the dependent variable. In natural resources, common covariates include elements of time (e.g., number of years since a disturbance to a forest or number of days a plant was subject to drought stress) or initial conditions prior to an event (e.g., the density of a forest prior a disturbance or the size of a plant prior to being exposed to drought stress). In ANCOVA, the covariate can be treated as it might be in a regression equation:

The ANCOVA model takes the form:

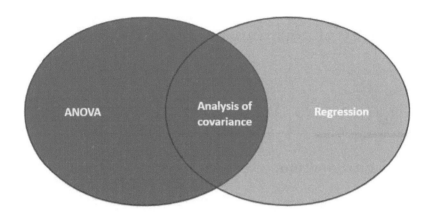

FIGURE 11.1 Analysis of covariance blends analysis of variance and regression.

$$y_{ij} = \mu + \tau_i + \mathbf{B}(x_{ij} - \bar{x}_i) + \varepsilon_{ij}$$

where,

- y_{ij} is the jth observation mean from the ith level,
- μ is the overall mean,
- τ_i is the effect of the ith level of the treatment,
- x_{ij} is the jth observation of the covariate in the ith level,
- $\bar{x}_i$ is the ith group mean, and
- ε_{ij} is the random error term.

Like any statistical test, there are several key assumptions that underlie the use of ANCOVA and affect the interpretation of the results. The assumptions used in linear regression and ANOVA also hold for ANCOVA analyses, i.e., the treatment effect and the covariate are independent, the variance is homogeneous, there is a linear relationship between the covariate and response variable, and residuals (errors) are normally distributed. For ANCOVA, an added assumption is that the **regression slopes are the same** for all treatment groups.

11.2.1 Testing equality of slopes in ANCOVA

If an assumption is that the slope of the covariate is equal across all treatment groups, it needs to be assessed prior to reporting the results from an ANCOVA. After plotting the data using scatter plots and line graphs, you can visually compare if slopes are equal:

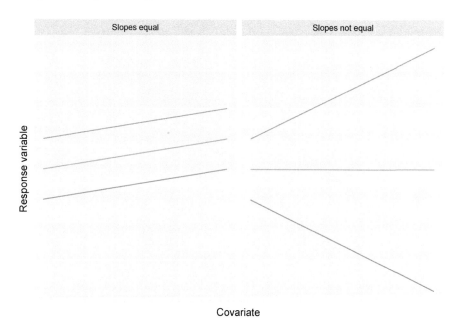

FIGURE 11.2 An assumption of the analysis of covariance is the equality of slopes across the range of values for the covariate.

As an example, consider the **fish** data set from the **stats4nr** package (Puranen 1917). Here, we can consider the species of fish as a factor and the length of the fish (`Length1`) as a covariate. We might be interested in predicting the weight of a fish (`Weight`) using these two variables. For simplicity, we'll create a data set that contains observations for two species: bream and pike:

```
library(stats4nr)

fish <- fish %>%
    filter(Species %in% c("Bream", "Pike"))
```

We can plot the data to see the weight-length differences between the two species:

```
p.fish <- ggplot(fish, aes(Length1, Weight, col = Species)) +
  geom_point() +
  stat_smooth(method = "lm")
p.fish
```

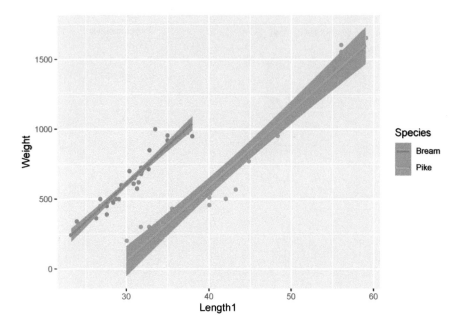

Can we make the claim that the two slopes are equal? Note that although the range of length measurements for pike is greater than for bream, for a given length of a fish, bream will weigh more than pike. For the ANCOVA analysis, our key interest lies in assessing the slopes of the two lines.

An appropriate statistical test to examine the equality of slopes is a *t*-test. The test statistic can be computed as:

$$t = \frac{\beta_{1,\,\text{bream}} - \beta_{1,\,\text{pike}}}{\sqrt{(SE_{bream}^2 + SE_{pike}^2)}}$$

where $\beta_{1,i}$ is the slope regression coefficient and SE_i is the standard error for the slope for each species.

We can fit two regressions to the bream and pike data simultaneously using a `group_by` statement. Then, the `tidy()` function from the **broom** package can return the regression coefficients and standard errors:

```
library(broom)

fish_spp <- group_by(fish, Species)

fish.coef <- do(fish_spp,
    tidy(lm(Weight ~ Length1, data = .)))

fish.coef
```

```
## # A tibble: 4 x 6
## # Groups:   Species [2]
##    Species term          estimate std.error statistic  p.value
##    <chr>   <chr>            <dbl>     <dbl>     <dbl>    <dbl>
## 1 Bream   (Intercept)     -1015.     91.2     -11.1 1.54e-12
## 2 Bream   Length1           54.1      2.99     18.1 2.18e-18
## 3 Pike    (Intercept)     -1541.    144.      -10.7 2.04e- 8
## 4 Pike    Length1           53.2      3.32     16.0 7.65e-11
```

Note the slope estimate for bream is 54.1 and the slope estimate for pike is 53.2. With the null hypothesis $H_0 : \beta_{1,bream} = \beta_{1,pike}$ and the alternative hypothesis $H_1 : \beta_{1,bream} \neq \beta_{1,pike}$, the t-statistic can be computed as:

$$t = \frac{54.1 - 53.2}{\sqrt{2.99^2 + 3.32^2}} = \frac{0.90}{4.47} = 0.201$$

With such a small value of t, we can logically accept the null hypothesis and conclude that the two slopes are equal. Performing the t-test by hand would be an approach to use if all that was available were the values of the estimates and their standard errors. Another way to test if the slopes are equal in R is to following this process:

- Fit a "full" model with an interaction effect to examine if the effect of fish length on weight depends on species. (This is the definition of an interaction.)
- Fit a "reduced" model without an interaction effect.

For the fish data, the models can be written as:

```
fish.full <- lm(Weight ~ Length1 * Species, data = fish)

fish.reduced <- lm(Weight ~ Length1 + Species, data = fish)
```

Two R objects can be added to an `anova()` statement to examine the null
hypothesis that the slopes are equal:

```
anova(fish.full, fish.reduced)
```

```
## Analysis of Variance Table
##
## Model 1: Weight ~ Length1 * Species
## Model 2: Weight ~ Length1 + Species
##   Res.Df     RSS Df Sum of Sq      F Pr(>F)
## 1     47  340835
## 2     48  341108 -1   -273.33 0.0377 0.8469
```

Similar to what was observed when manually calculating the t-statistic, we
observe a large p-value (0.8469) comparing the two models, indicating the
effect of fish length on weight does not depend on species. Hence, we can
conclude with our data of bream and pike observations that we can perform
ANCOVA procedures because the slopes are equal for each species across the
range of the covariate `Length1`.

11.2.2 ANCOVA hypothesis tests and outcomes

Hypothesis tests for the ANCOVA model are very similar to an ANOVA,
but the key difference is that the population means for each treatment t are
adjusted for the covariate. For example, we can label the population mean of
the first factor level as μ_1^*. The null and alternative hypotheses can be written
as

$$H_0 = \mu_1^* = \mu_2^* = ... = \mu_t^*$$

$$H_A = \mu_i^* \neq \mu_j^*, \text{for some } i \neq j$$

The variance can be partitioned in an ANCOVA model by examining three
components:

- The **treatment effect** reflects the main treatment effect or factor in the
  study, e.g., the species of fish.

- The **covariate effect** adjusts the sums of squares for the covariate, i.e., the
  length of the fish.

- The **error** represents the sum of squares within each treatment group.

TABLE 11.1 The analysis of covariance (ANCOVA) table.

Source	Degrees of freedom	Sums of squares	Mean square	F
Covariate	1	SS(Cov)	MS(Cov) = SS(Cov)	MS(Cov) / MSW
Treatment	k-1	SS(B)	MSB = SSB / k-1	MSB / MSW
Error (Within)	n-k-1	SSW	MSW = SSW / n-k-1	-
Total	n-1	TSS	-	-

You can see both elements of ANOVA and regression in the ANCOVA model. The "regression" component is represented by the covariate and the "ANOVA" part is represented by the treatment. Of course, there will also be error in our model that cannot be explained by the treatment or covariate.

We can use the ANCOVA table to make inference about the factors of interest and the covariate. Note how we can carry out two hypothesis tests with ANCOVA: one for the covariate and one for the treatment effect (as noted in the two *F*-statistics we calculate).

Similar to ANOVA, multiple comparisons can be performed on the treatment effect within an ANCOVA model. Given we already know there are differences between the weights of bream and pike fish, the LSD.test() function from the

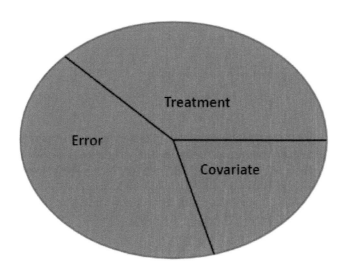

FIGURE 11.3 Variance partitioning in analysis of covariance.

agricolae package provides different letters denoting significant differences between the two species:

```
library(agricolae)

lsd.fish <- LSD.test(fish.reduced, "Species", p.adj = "bonferroni")
lsd.fish
```

```
## $statistics
##    MSerror Df    Mean      CV
##   7106.423 48 656.902 12.8329
##
## $parameters
##          test  p.ajusted  name.t ntr alpha
##   Fisher-LSD bonferroni Species   2  0.05
##
## $means
##         Weight      std  r      LCL      UCL Min  Max    Q25 Q50   Q75
## Bream 626.0000 206.6046 34 596.9317 655.0683 242 1000 481.25 615 718.5
## Pike  718.7059 494.1408 17 677.5971 759.8146 200 1650 345.00 510 950.0
##
## $comparison
## NULL
##
## $groups
##         Weight groups
## Pike   718.7059      a
## Bream  626.0000      b
##
## attr(,"class")
## [1] "group"
```

With a mean of 719 and 626 grams for pike and bream, respectively, we find there are significant differences between the weights of the two species after adjusting for the covariate Length1. In terms of the consequences of outcomes, the addition of the covariate in ANCOVA reduces the probability of Type II error, i.e., failing to reject the null hypothesis when the null hypothesis is false.

11.2.3   Exercises

11.1 Make a bar plot showing the fish weight results presented earlier in this chapter.

    a. Write code in `ggplot()` that displays two bars that show the mean fish weight for bream and pike. For error bars, present the 95% confidence limits for mean weights. (HINT: Assume a value t of 1.96 in your calculation.) Add letters denoting significant differences above the bar plots between the two species.

    b. What interesting finding do you observe comparing the bar plot you created with the ANOVA output?

11.2 Perform ANCOVA on the entire **fish** data set, i.e., on all seven species and 157 observations. Using these data, evaluate whether or not you meet the assumptions of ANCOVA and report your results.

11.3 Perform an ANCOVA using **Loblolly**, a data set built into R. The data set contains the height, age, and seed source of loblolly pine (*Pinus taeda*) trees.

    a. Create a scatter plot of `age` (*x*-axis) and `height` (*y*-axis), adding color and a linear regression line (without confidence intervals) for each seed source (`Seed`).

    b. Write code to evaluate the assumptions of whether or not an AN-COVA can be performed to determine loblolly pine tree height using seed source as a factor and age as a covariate.

    c. Summarize the ANCOVA results using a statistical test at $\alpha = 0.10$. Do specific seed sources produce taller heights of loblolly pine trees?

11.3   Summary

Analysis of covariance blends components of linear regression and analysis of variance by integrating a quantitative covariate and a categorical treatment variable. The primary difference between an ANOVA and an ANCOVA model is the addition of the continuous covariate. An analyst should select an appropriate covariate that explains a portion of the variability in the dependent variable of interest. A key step in performing an ANCOVA analysis is to first determine the equality of slopes when investigating the covariate-response variable relationship at different levels of the treatment or factor variable. The ANCOVA adjusts each treatment mean based on the dependent variable.

FIGURE 11.4 An old-growth loblolly pine tree. Image: Wikimedia Commons

You may encounter obstacles when performing an ANCOVA. For example, what if the slopes are examined and the results are that they are not equal? In this case, you have three options for moving forward with your analysis. First, consider dropping the covariate and running a one-way ANOVA. Second, consider "categorizing" the covariate and running a two-way ANOVA. As an example, if slopes were not equal in the fish data set, we might have created an indicator variable that labeled them as long and short fish, using a threshold value for length that we deem appropriate. Lastly, you could retain the covariate and consider fitting a general linear model. In the event that you have several covariates that are a part of a data set, consider multiple regression techniques.

11.4   Reference

Puranen, J. 1917. Fish catch data set. *Journal of Statistics Education, Data Archive*: Available at: http://jse.amstat.org/datasets/fishcatch.txt

12

Logistic regression

12.1 Introduction

One of the primary assumptions of the regression techniques we have learned up to this point is that the response variables are independent normal random variables, i.e., the normality assumption. Similarly, linear models using ordinary least squares regression are not appropriate when the response y_i is stored as a binary or count variable or when the variance of y_i depends on the mean. For example, the relationship between the size of an organism and its age is curved, i.e., growth tends to rise in juvenile stages, flatten out, and possibly decline after that.

This chapter will discuss how generalized linear models can be used to perform logistic regression (when the response variable is a binary variable), multinomial logistic regression (when there are three or more categorical variables for the response variable), and ordinal regression (when there are three or more categorical variables for the response variable and they are ordered or ranked in a meaningful way).

12.2 An overview of generalized linear models

Generalized linear models (GLMs) extend the linear modeling framework to response variables that are not normally distributed. The GLM framework generalizes linear regression by allowing the linear model to be related to the response variable by using a link function. The link function describes the relationship between the linear predictor and its mean value, while a variance function describes the relationship between the variance and mean. A GLM model can be broadly written as

$$\eta_i = \beta_0 + \beta_1 x_{i1} + \beta_2 x_{i2} + \dots + \beta_k x_{ik}$$

TABLE 12.1 Common link functions used in generalized linear models with natural resources data.

Family	Link function	Data	Example
Gaussian	Identity	Normally distributed data	Height of a plant
Binomial	Logit	A binary or multinomial variable	Whether a plant is alive or dead
Poisson	Log	Count data (e.g., 0, 1, 2, 3)	Number of plants in a square meter plot

where η_i represents the linear predictor, β_i represent parameters, and x_i represent independent variables.

The appropriate link function to specify relies on how best to describe the error distribution in the model. Three common link functions used with natural resources data include the Gaussian, binomial, and Poisson types. The Gaussian link function is most similar to linear regression techniques where data are distributed normally with a range of negative and positive values. The binomial link function is used when the response variable is categorical, e.g., a binary or multinomial variable. A binomial count could also consist of the number of "yes" occurrences out of N "yes/no" occurrences. A Poisson link function is used with count data where the response variable is stored as an integer, e.g., 0, 1, 2, 3, etc.

In R, the `glm()` function fits GLMs. The function is similar to `lm()` which is used in regression and analysis of variance. The main difference is that we need to include an additional argument that describes the error distribution and link function to be used in the model. In the `family =` statement within `glm()`, the link functions for the Gaussian, binomial, and Poisson types can be specified as `link = identity`, `link = logit`, and `link = log`, respectively.

12.2.1 Logistic regression

Logistic regression is one of the most common types of GLMs. Logistic regression analyzes a data set in which there are one or more independent variables that determine an outcome. This differs greatly from linear regression because in logistic regression, the dependent variable is categorical. Hence, logistic regression models a binomial count, e.g., the number of "yes" or "no" occurrences.

TABLE 12.2 Relationship between probability and odds ratios.

Probability	Odds	Log odds
0.05	0.05	-3.00
0.25	0.33	-1.11
0.50	1.00	0.00
0.75	3.00	1.10
0.95	19.00	2.94

Concepts presented in Chapters 5 and 6 deal with hypothesis tests of count and proportional data, however, logistic regression will allow you to model the probability of a binomial count occurring with a suite of predictors. In natural resources, logistic regression is often used to quantify the presence/absence of some phenomenon or the living status of an organism (e.g., alive or dead).

We learned that the logit link function can be used to model a binary variable. A logistic regression model is written as:

$$\text{logit}(p) = \beta_0 + \beta_1 x_1 + \beta_2 x_2 + ... + \beta_k x_k$$

The logit function relies on knowing the probability of success p and dividing it by its complement $1 - p$, or the probability of failure. The odds ratio is often presented as a log odds after taking its logarithm. In doing this, log odds have values centered around zero:

$$\text{logit}(p) = \log \frac{p}{1 - p}$$

For example, consider that 65% of trees in a forest were defoliated by an insect. The odds ratio would be $0.65/0.35 = 1.85$ and a log odds ratio of $\log(0.65/0.35) = 0.62$. Values with negative log odds are associated with probabilities of success less than 0.50 and values with positive log odds are associated with probabilities of success greater than 0.50.

Logistic regression is fit with maximum likelihood techniques using one or more independent variables. Similar to how we can evaluate multiple regression models, the Akaike's information criterion (AIC) can be used to evaluate the quality of logistic regression models. The AIC is calculated as

$$\text{AIC} = 2k - 2ln(\hat{L})$$

where L represents the maximum value for the likelihood of the model and k represents the number of parameters in the model. The AIC value can be compared across different models fit to the same dependent variable. A lower AIC for a model indicates a higher-quality model if a difference of two or more units is observed.

FIGURE 12.1 Tree seedling showing evidence of continual browse damage from white-tailed deer. Photo by the author.

12.2.1.1 Case study: Predicting deer browse on tree seedlings

To see how logistic regression is implemented in R, an experiment that planted oak (*Quercus*) trees in fenced and non-fenced areas was conducted at the Cloquet Forestry Center in Cloquet, MN. In this region, oak seedlings are palatable to white-tailed deer (*Odocoileus virginianus*) that use them as a food source. In addition to fencing, oaks were planted close together to minimize risk from deer browse. This experiment was based on the German "Truppflanzungs" idea of the 1970s (Saha et al. 2017). Oak seedlings were planted in tight clusters with 20 to 30 individuals per square meter. The goal of the experiment was to provide conditions so that at least one tree per cluster would survive and thrive in forests subject to severe herbivory pressure from white-tailed deer.

The data are found in the **trupp** data set in the **stats4nr** R package. There are nine variables and 762 observations in the data:

```
library(stats4nr)

trupp
```

```
## # A tibble: 762 x 9
##     plot   trmt   tree  year    dia    ht browse   dead missing
##     <chr>  <chr> <dbl> <dbl>  <dbl> <dbl>  <dbl>  <dbl>   <dbl>
##  1 P1     FENCE     1  2007    0       0      0      1       0
##  2 P1     FENCE     2  2007    8.7    47      0      0       0
##  3 P1     FENCE     3  2007    6.7    65      0      0       0
##  4 P1     FENCE     4  2007    7.6    41      0      0       0
##  5 P1     FENCE     5  2007    6.2    26      0      0       0
##  6 P1     FENCE     6  2007    4.9    33      0      0       0
##  7 P1     FENCE     7  2007    9.2    46      0      0       0
##  8 P1     FENCE     8  2007    8.3    45      0      0       0
##  9 P1     FENCE     9  2007    6.3    46      0      0       0
## 10 P1     FENCE    10  2007    8.5    62      0      0       0
## # ... with 752 more rows
```

The variables are:

- `plot`: Plot identification number
- `trmt`: Fencing treatment (FENCE, NOFENCE, CTRL)
- `tree`: Tree identification number
- `year`: Year of measurement (2007, 2008, or 2013)
- `dia`: Seedling diameter (mm)
- `ht`: Seedling height (cm)
- `browse`: Indicator for whether or not seedling was browsed (1) or not (0)
- `dead`: Indicator for whether or not seedling was dead (1) or not (0)
- `missing`: Indicator for whether or not seedling was missing (1) or not (0)

Our interest is in predicting whether or not an oak seedling was browsed based on its diameter and whether it was located in the fenced treatment. Prior to performing the logistic regression, we will remove the control plots to directly compare the fenced and non-fenced treatments:

```
trupp <- trupp %>%
  filter(trmt != "CTRL")
```

The data result in 522 observations. To visualize the results, we can see that seedlings in fenced areas tend to be browsed less (`browse = 0`) and are larger in diameter than non-fenced seedlings:

```
p.browse <- ggplot(trupp, aes(dia, browse, col = trmt)) +
  geom_point() +
```

```
    labs(x= "Diameter (mm)", y = "P(browse)")
p.browse
```

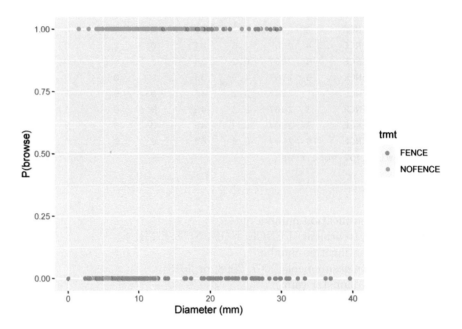

The `glm()` function in R is used to specify a logistic regression. We model the response variable `browse` using the predictor variables `dia` and `trmt`. We specify the logit link function within the `family` statement to indicate a binary response in our model:

```
browse.m1 <- glm(browse ~ dia + trmt,
                 family = binomial(link = 'logit'),
                 data = trupp)
summary(browse.m1)
```

```
##
## Call:
## glm(formula = browse ~ dia + trmt, family = binomial(link = "logit"),
##     data = trupp)
##
## Deviance Residuals:
##     Min       1Q   Median       3Q      Max
## -2.1313  -0.7953  -0.1212   0.5361   3.0946
##
```

```
## Coefficients:
##               Estimate Std. Error z value Pr(>|z|)
## (Intercept) -6.37547    0.69352  -9.193  < 2e-16 ***
## dia          0.21560    0.02696   7.998 1.26e-15 ***
## trmtNOFENCE  4.77196    0.53263   8.959  < 2e-16 ***
## ---
## Signif. codes:  0 '***' 0.001 '**' 0.01 '*' 0.05 '.' 0.1 ' ' 1
##
## (Dispersion parameter for binomial family taken to be 1)
##
##     Null deviance: 645.02  on 521  degrees of freedom
## Residual deviance: 389.98  on 519  degrees of freedom
## AIC: 395.98
##
## Number of Fisher Scoring iterations: 6
```

The summary output provides the deviance residuals, coefficients, and AIC value for the logistic model. The residual deviance measure is analogous to the residual sums of squares metric used in linear regression. The predictor variables dia and trmt are statistically significant and their coefficients provide the change in the log odds of the browse. For interpretation, for every one-millimeter change in dia, the log odds of browse increases by 0.22. For seedlings in non-fenced areas, changes the log odds of browse increases by 4.77.

The log odds provided in R output for a logistic regression model is helpful but not easily interpreted. You can exponentiate the coefficients to determine the odds ratios directly:

```
exp(coef(browse.m1))
```

```
##   (Intercept)          dia   trmtNOFENCE
## 1.702813e-03 1.240605e+00 1.181504e+02
```

So, for a one-unit increase in dia, the odds of an oak seedling being browsed increases by a factor of 1.24. For seedlings in non-fenced areas, the odds of them being browsed increases by a factor of 118. The conclusions of the experiment are that seedlings with a larger diameter that are found in non-fenced areas have a greater probability of being browsed.

12.2.1.2 Logistic regression from a 2x2 table

The **trupp** data set is formatted "long" such that each row contains a seedling observation. When dealing with binary outcomes in logistic regression, it is also common to record data in a 2x2 contingency table with the number of observations in each cell. This is especially useful when dealing with categorical predictor variables (e.g., trmt in the **trupp** data). However, applications with continuous predictor variables that vary for each observation (e.g., dia in the **trupp** data) require them to be organized in a long format for logistic regression.

For categorical variables, a logistic regression can be performed on data organized in a 2x2 contingency table. Data could be entered manually in this format (see Chapter 5 and the matrix() function) or an existing long data set can be modified into a 2x2 table using the with() and table() functions. For example, a new data set with trmt and browse as row and column name labels can be created from the **trupp** data set containing the number of observations in each category:

```
trupp.table <- with(trupp, table(trmt, browse))
trupp.table
```

```
##            browse
## trmt          0    1
##    FENCE    242   19
##    NOFENCE  119  142
```

To perform a logistic regression on this data set, adding the weights = Freq statement to the glm() function tells R to use the counts of observations in the model fitting:

```
browse.m2 <- glm(browse ~ trmt, weights = Freq,
                 family = binomial(link = 'logit'), data = trupp.table)
summary(browse.m2)
```

```
##
## Call:
## glm(formula = browse ~ trmt, family = binomial(link = "logit"),
##     data = trupp.table, weights = Freq)
##
## Deviance Residuals:
##       1        2        3        4
##  -6.048  -13.672    9.978   13.148
##
## Coefficients:
##              Estimate Std. Error z value Pr(>|z|)
## (Intercept)   -2.5445     0.2383  -10.68   <2e-16 ***
```

```
## trmtNOFENCE    2.7212      0.2687    10.13    <2e-16 ***
## ---
## Signif. codes:  0 '***' 0.001 '**' 0.01 '*' 0.05 '.' 0.1 ' ' 1
##
## (Dispersion parameter for binomial family taken to be 1)
##
##      Null deviance: 645.02  on 3  degrees of freedom
## Residual deviance: 495.94  on 2  degrees of freedom
## AIC: 499.94
##
## Number of Fisher Scoring iterations: 6
```

An interested learner might try fitting the same logistic regression model using the **trupp** data set, which would yield identical results compared to the browse.m2 model. Adding more categorical variables as predictors and including their interactions, it is possible to extend the logistic regression model. Then, the quality of multiple logistic regression models could be compared using AIC.

12.2.1.3 Visualizing results and making predictions

We can modify the p.browse figure by adding the logistic curves for fenced and non-fenced areas. In ggplot() the stat_smooth() statement can include GLM model types if the binomial family is specified:

```
p.browse +
  stat_smooth(method = "glm", se = FALSE,
              method.args = list(family = binomial))
```

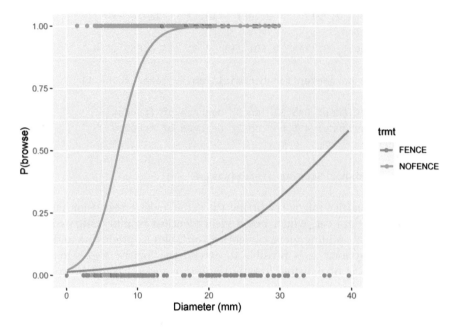

Here we can see the clear effects that fencing has in lowering the probability of deer browse on oak seedling. If we have new seedling observations and we seek to predict the probability of deer browse, the handy `predict()` function works well. For example, consider we have the following six seedlings and we want to estimate the probability of them being browsed:

```
new.seedling <- tibble(
  dia = c(10, 10, 20, 20, 30, 30),
  trmt = c("FENCE", "NOFENCE", "FENCE",
           "NOFENCE", "FENCE", "NOFENCE"))
```

We add a new variable termed `P_browse` which indicates the probability of a seedling being browsed. Specifying `type = "response"` provides the predicted probability:

```
new.seedling <- new.seedling %>%
  mutate(P_browse = predict(browse.m1,
                            newdata = new.seedling,
                            type = "response"))
new.seedling
```

```
## # A tibble: 6 x 3
##      dia trmt     P_browse
##    <dbl> <chr>       <dbl>
## 1    10 FENCE      0.0145
## 2    10 NOFENCE    0.635
## 3    20 FENCE      0.113
## 4    20 NOFENCE    0.938
## 5    30 FENCE      0.523
## 6    30 NOFENCE    0.992
```

12.2.1.4 Assessing accuracy with the confusion matrix

Accuracy and error are two key metrics when using logistic regression to classify data. You can interpret the performance of the model much more deeply if you learn more about another valuable tool: the **confusion matrix**.

The confusion matrix includes rows and columns for all possible labels in a classification:

- **True positives (TP)**: The model correctly predicts a positive value.
- **True negatives (TN)**: The model correctly predicts a negative value.
- **False positives (FP)**: The model incorrectly predicts a positive value.
- **False negatives (FN)**: The model incorrectly predicts a negative value.

Each cell in the confusion matrix contains the number of instances that are classified in a certain way. We're most interested in getting true positives and true negatives: this means our logistic model is working properly. A greater number of false positives and false negatives will decrease the accuracy of the model.

Using the confusion matrix, the accuracy can be calculated as the number of correct predictions divided by the total number of predictions made in the data set:

$$Accuracy = \frac{TP + TN}{TP + TN + FP + FN}$$

Using the confusion matrix, the accuracy and error (1 - accuracy) measure how well the logistic regression performs in classifying outcomes. While this method is often used on independent data not used in the fitting of the model, for our application we will create a confusion matrix for the seedling-browse data. We first start by making predictions of the probability of browse (P_browse) on the **trupp** data:

```
trupp <- trupp %>%
  mutate(P_browse = predict(browse.m1,
                            newdata = trupp,
                            type = "response"))
```

While logistic regression provides a range of probabilities for some event occurring, we need to establish a threshold probability, or cutoff probability, to
define when an event will occur. In our example, we will set a threshold probability of 0.50: if P_browse is greater than 0.50, we will label the seedling as
being browsed. Otherwise, we will label the seedling as not being browsed. In
R, we create factor variables containing yes/no categories to label the browse
predictions (browse_pred) and observations (browse_obs):

```
P_threshold = 0.50

trupp <- trupp %>%
  mutate(browse_pred = factor(ifelse(P_browse > P_threshold,
                                     "yes", "no")),
         browse_obs = factor(ifelse(browse == 1,
                                    "yes", "no")))
trupp
```

```
## # A tibble: 522 x 12
##    plot  trmt   tree  year   dia    ht browse  dead missing
##    <chr> <chr> <dbl> <dbl> <dbl> <dbl>  <dbl> <dbl>   <dbl>
##  1 P1    FENCE     1  2007   0       0      0     1       0
##  2 P1    FENCE     2  2007   8.7    47      0     0       0
##  3 P1    FENCE     3  2007   6.7    65      0     0       0
##  4 P1    FENCE     4  2007   7.6    41      0     0       0
##  5 P1    FENCE     5  2007   6.2    26      0     0       0
##  6 P1    FENCE     6  2007   4.9    33      0     0       0
##  7 P1    FENCE     7  2007   9.2    46      0     0       0
##  8 P1    FENCE     8  2007   8.3    45      0     0       0
##  9 P1    FENCE     9  2007   6.3    46      0     0       0
## 10 P1    FENCE    10  2007   8.5    62      0     0       0
## # ... with 512 more rows, and 3 more variables:
## #   P_browse <dbl>, browse_pred <fct>, browse_obs <fct>
```

We can then compute the confusion matrix manually through a series of R calculations. However, the **caret** package contains the confusionMatrix() function to calculate this for us. The **caret** package contains a number of functions
that help to create and evaluate classification and regression models.

```
install.packages("caret")
library(caret)
```

We pass the factor variables browse_pred and browse_obs to the confusionMa-trix() function:

```
confusionMatrix(data = trupp$browse_pred, reference = trupp$browse_obs)
```

```
## Confusion Matrix and Statistics
##
##           Reference
## Prediction  no  yes
##        no   333   64
##        yes   28   97
##
##               Accuracy : 0.8238
##                 95% CI : (0.7883, 0.8555)
##    No Information Rate : 0.6916
##    P-Value [Acc > NIR] : 4.317e-12
##
##                  Kappa : 0.5596
##
##  Mcnemar's Test P-Value : 0.0002633
##
##            Sensitivity : 0.9224
##            Specificity : 0.6025
##         Pos Pred Value : 0.8388
##         Neg Pred Value : 0.7760
##             Prevalence : 0.6916
##         Detection Rate : 0.6379
##   Detection Prevalence : 0.7605
##      Balanced Accuracy : 0.7625
##
##       'Positive' Class : no
##
```

The confusion matrix is printed first in the output, followed by a number of additional statistics. The accuracy of the model is 82.38%, which can be confirmed with the values from the confusion matrix:

$$Accuracy = \frac{(333 + 97)}{(333 + 97 + 64 + 28)} = 82.38$$

TABLE 12.3 State results of West Nile occurrence in ruffed grouse.

State	Number of grouse sampled	Number of grouse with West Nile	Proportion of grouse with West Nile
Michigan	213	28	0.13
Minnesota	273	34	0.12
Wisconsin	235	68	0.29

So, with 82% accuracy, the logistic regression model can predict deer browse on seedlings correctly about four out of every five times. Acceptable accuracy values with natural resources data tend to be greater than 80%, however, many plant and animal populations will show lower accuracy values depending on the specific characteristic being studied. Fitting new logistic regression models and evaluating other predictor variables may be required if you find your model is predicting too many false positives or false negatives.

12.2.2   Exercises

12.1 In a 2018 survey on invasive plant management (Reinhardt et al. 2019), public land professionals and private landowners reported on the effectiveness of different treatments for controlling common buckthorn, an invasive shrub:

- 72% of respondents said that herbicide treatments were effective.
- 50% of respondents said that manual removal treatments were effective.
- 23% of respondents said that mechanical removal treatments were effective.

Use R code to find odds ratios for the probability of effectiveness for these various treatments.

12.2 Perform a logistic regression using data on West Nile occurrence in ruffed grouse populations using the following data:

   a. Fit a logistic regression model that determines West Nile infection based on the state in which the grouse was sampled.
   b. Use R code to determine the odds ratios for each of the coefficients (i.e., the three states).
   c. Use the logistic regression output and the `predict()` function to determine the probability of infection for a grouse in each of the three states.

12.3 You hypothesize that the crown position of cedar elm trees may be related to their diameter and height. Using the **elm** data set from the **stats4nr** package, perform the following analyses:

a. Create a binary variable that categorizes a tree's crown class code (CROWN_CLASS_CD) as either being intermediate or suppressed, or something else. Create a new binary variable in the **elm** data set that creates this knowing that CROWN_CLASS_CD is 4 and 5 for intermediate and suppressed trees, respectively. Then, make two box plots using ggplot() that show the distribution of these categories across a range of tree diameter (DIA) and height (HT). What characteristics do trees with intermediate or suppressed crown classes have compared to others?

b. Fit a logistic regression model using DIA and HT that predicts whether or not a tree is intermediate or suppressed or not. Do the results shown by the coefficients in the output confirm what you observed in the box plots?

c. Use the predict() function to determine how well your logistic model predicts whether or not a tree is intermediate or suppressed or not. Use a threshold probability of 0.50 and create a new variable in the **elm** data set.

d. Evaluate the predictions made in step c by creating a confusion matrix. What is the accuracy of the fitted logistic model?

e. Change the threshold probability to 0.45 and create a new confusion matrix. What is the accuracy for these predictions and how does it compare to what you found in step d?

12.4 Consider we have a data set with tree diameters and heights from an even-aged forest plantation. We know whether or not these trees are alive or dead. Consider we use logistic regression to create a classification model that categorizes 16 different trees as being alive or dead. Short trees with small diameters are mostly categorized as not alive (dead), indicated by everything to the left of a "prediction" line. Large diameter trees that are tall are alive, indicated by everything to the right of the line:

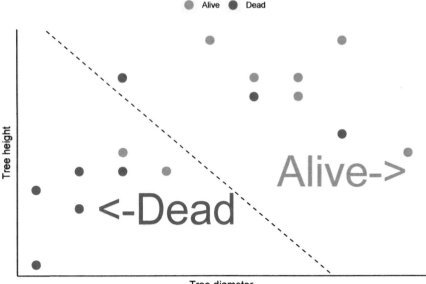

You can see that the model does a fairly good job of categorizing the trees as alive or dead. But there are three dead trees that were predicted to be alive. Plus, two alive trees were predicted to be dead.

- a. On paper or in R, create a small table that displays the confusion matrix for these observations and predictions. Label the true positive, false negative, false positive, and true negative values.
- b. Determine the accuracy of this model.
- c. The **precision** of a classification model measures the proportion of positive classifications that are correct. It is calculated as the number of true positive predictions divided by the total number of positive predictions (true and false). Calculate the precision of this model.
- d. The **sensitivity** (or recall) of a classification model measures the proportion of correct positives that were identified correctly. It is calculated as the number of true positive predictions divided by the total number of true positive and true negative predictions. Calculate the sensitivity of this model.

12.3 Modeling more than two responses

12.3.1 Multinomial logistic regression

There are many applications in natural resources when there is a binary response. But many other cases involve situations when there are three or more responses. In this case, a **multinomial logistic regression** can be performed when the response variable contains three or more categories that are not ordered. For example, a multinomial logistic regression can be performed to predict the nesting locations for a data set of bird species based on their physiological characteristics. These response variables could include trees, shrubs, the ground, or cliff and other high open areas.

Multinomial models are similarly fit using maximum likelihood techniques with a multinomial distribution. As the number of responses increases, the multinomial model becomes more complex. In R, the **nnet** package is useful for fitting multinomial logistic regression models. The package was created for designing neural networks, a tool used widely in machine learning (Venables and Ripley 2002):

```
install.packages("nnet")

library(nnet)
```

To see how logit regression models are fit to data with more than two responses, we will use data collected from a survey of over 300 private forest landowners in Minnesota, USA (UMN Extension Forestry 2021). The survey conducted an educational program needs assessment for this audience, inquiring about topics that participants are interested in learning about and future educational needs. Specifically, we may be interested in how differences in the ages of learners within this audience influence their preferred formats for learning about new educational opportunities and their comfort in virtual learning.

One question in the needs assessment inquired about the preferred format to learn about upcoming educational programs offered through the university. Landowners chose from the options of email, printed newsletter, social media, or website or blog. The data were collected across a range of four age groups:

FIGURE 12.2 Forest landowners take part in an educational program. Photo by the author.

```
m1.table
```

```
##             method
## age         Email Printed newsletter Social media Website or blog
##    18-34        8                   0            0               1
##    35-54       38                   2            0               5
##    55-69       91                   7            0               2
##    70 or over  43                  10            1               0
```

You will note that most landowners preferred email as a format, with fewer responses in the other categories. We seek to determine the preferred format based on age group because the response variable is not ordered and there are three or more categories. The `multinom()` function from the **nnet** package fits multinomial logistic regressions in a format similar to logistic regression. Again, specifying `weights = Freq` indicates to use the number of observations contained in the `m1.table` contingency table:

```
multi.learn <- multinom(method ~ age, weights = Freq, data = m1.table)
```

```
## # weights:  20 (12 variable)
## initial  value 288.349227
## iter  10 value 104.896727
## iter  20 value 92.578146
## iter  30 value 92.467996
## iter  40 value 92.448798
## iter  50 value 92.446478
```

```
## final   value 92.446474
## converged
```

```
summary(multi.learn)
```

```
## Call:
## multinom(formula = method ~ age, data = m1.table, weights = Freq)
##
## Coefficients:
##                      (Intercept)     age35-54     age55-69 age70 or over
## Printed newsletter   -16.153156 13.20863573   13.588153      14.694491
## Social media         -11.222075 -5.36141505  -11.896637       7.460595
## Website or blog       -2.079573  0.05147451   -1.738202     -17.915025
##
## Std. Errors:
##                      (Intercept)     age35-54     age55-69 age70 or over
## Printed newsletter    0.2240912    0.5598176   0.35657182   3.344345e-01
## Social media         96.6680976  654.5700991   0.05515755   9.667339e+01
## Website or blog       1.0607205    1.1625120   1.27911905   8.284796e-06
##
## Residual Deviance: 184.8929
## AIC: 208.8929
```

The output reports the odds of selecting one of the preferred learning formats as a reference scenario (in this case, email). To decipher the output, the coefficients and standard errors for the numbers in the Printed newsletter row compare models determining the probability of preferring printed newsletters as a method compared to email. Values in other rows similarly compare different preferred methods relative to the reference email format. As an example, the log odds of preferring a printed newsletter compared to an email will increase by 14.69 if comparing those in the 18-34 to the 70 years or over age group.

Similar to logistic regression, you can exponentiate the coefficients in a multinomial logistic regression to determine the odds ratios:

```
exp(coef(multi.learn))
```

```
##                      (Intercept)     age35-54     age55-69 age70 or over
## Printed newsletter 9.655467e-08 5.450516e+05 7.966356e+05  2.408442e+06
## Social media       1.337564e-05 4.694259e-03 6.813277e-06  1.738182e+03
## Website or blog    1.249836e-01 1.052822e+00 1.758363e-01  1.658072e-08
```

The predict() function can be used estimate multinomial probabilities. We will create a simple data set named **age** containing the four age groups, then bind it to a data set of predicted values containing the probabilities determined from the multi.learn object. We will then apply the pivot_longer() function to create a tidy data set containing three variables: age group, learning format, and probability.

```r
age <- tibble(age = c("18-34", "35-54", "55-69", "70 or over"))

pred_multi <- cbind(age, predict(multi.learn,
                                 newdata = age,
                                 type = "probs"))

pred_multi_long <- pred_multi %>%
  pivot_longer(!age, names_to = "Format", values_to = "Prob")
```

We can plot the data to visualize the trends across age groups and the preferred format of learning about educational programs. A bar graph that is dodged by preferred format highlights these trends:

```r
ggplot(pred_multi_long, aes(x = age, y = Prob, fill = Format)) +
  geom_bar(stat = "identity", position = "dodge")
```

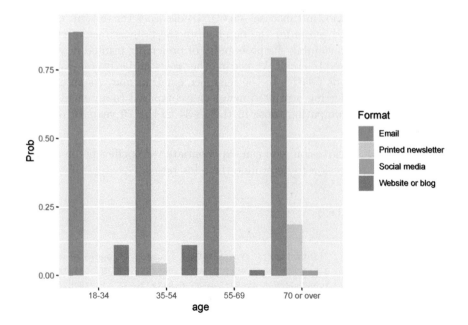

Here we can see the greater probability of older participants preferring printed newsletters as a format for learning about new programs. While younger audiences favor email heavily, older individuals have a lower preference for email as a medium.

DATA ANALYSIS TIP: Oftentimes it may be appropriate to "dichotomize" a multinomial variable into a binary one so that a simpler logistic regression analysis can be performed. For example, we could turn the learning format method variable into a binary one by labeling the formats as *print* (e.g., the print newsletter responses) or *online* (e.g., the email, social media, and website/blog responses). Then, we could perform logistic regression on this binary variable. Use caution with this approach, as you may lose important insights from specific groups when you combine and code them into different categories.

12.3.2 Ordinal regression

There may be other cases when there are three or more responses and there is an ordered relationship between the response variables. If a categorical variable is **ordinal** it indicates that the order of values is important, but the difference between each order cannot be quantified. A few examples in natural resources include:

- A survey provides a statement to participants and asks them to evaluate it on a five-point scale (e.g., a Likert (1932) scale) consisting of the responses strongly disagree, disagree, neither agree nor disagree, agree, or strongly agree.
- Pieces of downed woody debris are measured for their decay class (Harmon et al. 1986), a somewhat subjective measure of the reduction in volume, biomass, carbon, and density of dead wood, typically labeled from one (freshly fallen) through five (advanced decomposition).
- The production of leaves following herbicide applications on bald cypress trees measured from none to severe malformation (Sartain and Mudge 2018).

In these cases, **ordinal regression**, also termed cumulative logit modeling, can be performed when the response variable contains three or more categories that are ordered. In R, the **MASS** package introduced by Venables and Ripley (2002) is useful for fitting ordinal regression models:

```
install.packages("MASS")

library(MASS)
```

As an example, a different question in the forestry programming needs assessment asked respondents about their comfort in participating in online learning. This was essential information to be aware of during the COVID-19 pandemic when the majority of education occurred in an online virtual environment. Respondents were provided the statement "I feel comfortable participating in online learning" and were asked to choose from the options of "Always," "Most of the time," or "Never." These responses can be considered an ordinal response because they reflect a decreasing comfort in online learning.

The data were summarized across the four age groups:

```
m2.table
```

```
##               comfort
## age            Always Most of the time Never
##    18-34           11                0     0
##    35-54           41                7     2
##    55-69           87               23     3
##    70 or over      36               28     6
```

Most respondents were comfortable with learning in an online environment at least most of the time, however, some individuals in older age groups indicated they were never comfortable with online learning. (As an aside, this was an interesting finding given that the needs assessment survey was an online one. Hence, the true number of individuals that are not comfortable with online learning is likely underestimated.)

In ordinal regression, the log odds of being less than or equal to a chosen category can be defined as:

$$log\frac{P(y \leq j)}{P(y > j)}$$

where y is an ordered response and j is the number of responses. Given the nature of the ordered variable, we will model comfort as the response variable using the polr() function from the **MASS** package, which indicates probit or ordered logistic regression. We follow the general approach as used in multinomial logistic regression when applying it to the m2.table contingency table:

```
ord.learn <- polr(comfort ~ age, weights = Freq, data = m2.table)
summary(ord.learn)
```

```
## Call:
## polr(formula = comfort ~ age, data = m2.table, weights = Freq)
##
## Coefficients:
##                   Value Std. Error t value
## age35-54          15.71     0.2875   54.63
## age55-69          15.99     0.1991   80.29
## age70 or over     17.13     0.2057   83.28
##
## Intercepts:
##                              Value    Std. Error t value
## Always|Most of the time    17.2017     0.1231   139.7747
## Most of the time|Never     19.4413     0.3095    62.8076
##
## Residual Deviance: 326.5716
## AIC: 336.5716
```

The output provides the coefficients associated with each age group in addition to the intercept values. The residual deviance and AIC values allow us to compare the quality of the models to others. Similarly, the predict() function can be run on an ordinal regression object:

```
pred_ord <- cbind(age, predict(ord.learn,
                        newdata = age,
                        type = "probs"))

pred_ord_long <- pred_ord %>%
  pivot_longer(!age, names_to = "Comfort", values_to = "Prob")
```

We can plot the predictions to visualize the trends across age groups and participant comfort in online learning:

```
ggplot(pred_ord_long, aes(x = age, y = Prob, fill = Comfort)) +
  geom_bar(stat = "identity", position = "dodge")
```

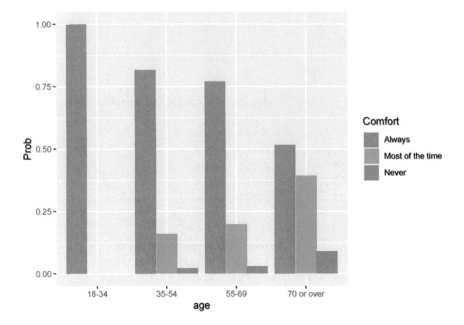

We see a greater probability of younger participants being comfortable with online learning, while older participants have a greater likelihood of stating they are comfortable with online learning "most of the time" or "never." Similar to multinomial logistic regression, you can exponentiate the coefficients in an ordinal regression to determine the odds ratios.

12.3.3   Exercises

12.5 After investigating the **fish** data set from the **stats4nr** package, you suspect that you might be able to correctly identify a species of fish based on only a few key measurements. Perform the following analyses:

    a. Create a scatter plot of `Width` and `Length1` and color each point to correspond to one of the seven different fish species.

    b. Using the `multinom()` function from the **nnet** package, create a multinomial logistic regression model that predicts fish species based on `Width` and `Length1`.

    c. Use the model to predict the fish species from the data set named **new.fish**, containing various fish measurements. HINT: add `type = "class"` within the `predict()` function obtain the name of the fish species:

```
new.fish <- tibble(Width = c(10, 10, 12, 15, 20),
                   Length1 = c(10, 40, 32, 22, 30))
```

12.6 In question 12.3, we created a binary variable that categorized a cedar elm's crown class code (`CROWN_CLASS_CD`) as either being intermediate or suppressed, or something else. We can actually consider the `CROWN_CLASS_CD` variable in the **elm** data set to be an ordinal one because trees that receive the most sunlight are categorized as a 1 (open-grown trees) and trees that receive the least sunlight are categorized as a 5 (suppressed trees).

    a. Perform an ordinal regression predicting `CROWN_CLASS_CD` using `DIA` and `HT` with the `polr()` function from the **MASS** package. The `CROWN_CLASS_CD` is stored as a numeric variable in the data, so remember to convert it to a factor level prior to modeling.

    b. Use the model to predict the probabilities of a tree being in each of the crown class codes from the data set named **new.elm**, containing various measurements of cedar elm trees. Which tree (i.e., its height and diameter) would have the greatest probability of being open grown? Which tree would have the greatest probability of being suppressed?

```
new.elm <- tibble(DIA = c(5, 10, 15, 20, 25, 30),
                  HT = c(20, 25, 30, 35, 40, 50))
```

12.4 Summary

Logistic regression has a number of applications in natural resources and is a useful tool when the response variable of interest and is categorical. Depending on the nature of the categorical response variable, the following provides a sample of which techniques and functions to use when fitting these models in R:

- Logistic regression is appropriate when the response variable is binary. Use the `glm()` function available in base R.
- Multinomial regression is appropriate when there are three or more categories. Use the `multinom()` function from the **nnet** package.
- Ordinal regression is appropriate when there are three or more categories

that are ordered in their design. Use the `polr()` function from the **MASS** package.

All of these functions are able to perform logistic regression procedures on a "tidy" data set of observations, where one row represents one observation, or a contingency table where categorical responses are summarized in a compact form. While most standard output in R and other statistical software programs provide coefficients in terms of log odds, exponentiating these values provides odds ratios which are more easily interpreted. Understanding both the confusion matrix and its accuracy values provides tremendous insight into the performance of logistic regression models, particularly when applied to data not used in model fitting. When comparing different models and the variables used in them, goodness of fit measures such as AIC work well in the evaluation process.

12.5　References

Harmon, M.E., J.F. Franklin, F.J. Swanson, P. Sollins, S.V. Gregory, J.D. Lattin, N.H. Anderson, S.P. Cline, N.G. Aumen, J.R. Sedell, G.W. Lienkaemper, K. Cromack, and K.W. Cummins. 1986. Ecology of coarse woody debris in temperate ecosystems. *Advances in Ecological Research* 15: 133–302.

Likert, R. 1932. A technique for the measurement of attitudes. *Archives of Psychology* 140: 1–55.

Reinhardt, J., M.B. Russell, W. Lazarus, M. Chandler, and S. Senay. 2019. Status of invasive plants and management techniques in Minnesota: results from a 2018 survey. University of Minnesota. Retrieved from the University of Minnesota Digital Conservancy. Available at: `https://hdl.handle.net/11299/216295`.

Saha, S., C. Kuehne, and J. Bauhus. 2017. Lessons learned from oak cluster planting trials in central Europe. *Canadian Journal of Forest Research* 47: 139–148.

Sartain, B.T., and C.R. Mudge. 2018. Effect of winter herbicide applications on bald cypress (*Taxodium distichum*) and giant salvinia (*Salvinia molesta*). *Invasive Plant Science and Management* 11: 136–142.

University of Minnesota Extension Forestry. 2021. Educational needs assessment of tree and woodland programs in Minnesota: results from a 2020 study. Retrieved from the University of Minnesota Digital Conservancy. Available at: `https://hdl.handle.net/11299/218206`

Venables, W. N., and B.D. Ripley. 2002. *Modern applied statistics with S. Fourth edition.* Springer, New York.

13

Count regression

13.1 Introduction

Count data are widely collected and analyzed across natural resources disciplines, particularly in wildlife biology and conservation. Count data represent values from non-negative integers, e.g., 0, 1, 2, 3, etc. Values are obtained from counts rather than ranks such as in ordinal regression. Examples include the number of hazardous waste violations issued by an agency to a company, the counts of birds by ornithologists, and the number of butterfly species observed by a prairie researcher.

Models of count data require special consideration compared to other regression techniques presented in this book. Natural resources professionals also collect data on variables that exhibit **zero-inflation (ZI)**, defined as an excess proportion of zeros relative to what would be expected under a given distribution. Count distributions such as the Poisson and negative binomial are useful in describing the stochastic nature of many variables. To quantify both structural and stochastic sources of variability, ZI count models are commonly employed.

13.2 Count data and their distributions

Count data represent discrete random variables given the nature of the data. The Poisson, binomial, and negative binomial distributions are commonly used distributions to reflect count data. The **Poisson** model is the benchmark model for count data, but it can be restrictive when estimating attributes other than the mean (Winkelmann 2008). The Poisson model is useful for describing events that are rare, independent, and collected from large populations. Its distribution is written as:

$$f_P(x) = \frac{\lambda^x e^{-\lambda}}{x!}$$

where x denotes a count of the number of occurrences (i.e., $x = 0, 1, 2...$) and λ represents the mean and variance of the distribution.

Negative binomial models are count models that are similar to Poisson models but include a dispersion parameter, making them more flexible and suitable for estimating a wide variety of count data. As this dispersion parameter increases, the variance of the distribution approaches the mean, making it similar to a Poisson distribution. In short, if the mean and variance are not equal to one another, dispersion is present, making for a useful case for employing the negative binomial distribution. A negative binomial probability can be obtained as a gamma mixture of Poisson distributions. If we consider x failures and r successes, the negative binomial is written as:

$$f_{NB}(x) = \binom{x + r - 1}{r - 1} p^r (1 - p)^x$$

where p indicates the probability of success. Compared to the normal, these count distributions have a different appearance for the same set of parameters:

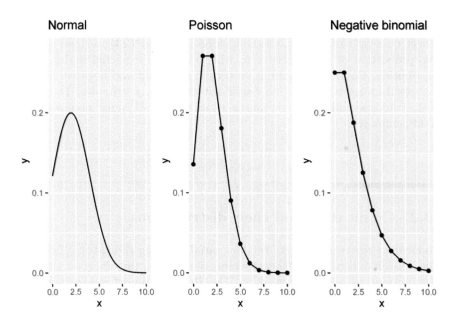

FIGURE 13.1 The normal, Poisson, and negative binomial distributions with a mean of 2 and variance of 2. (Dispersion parameter set to 0.5 for negative binomial.)

13.2.1 Case study: predicting fishing success

The **fishing** data set in the **stats4nr** package contains data on the number of fish caught by visitors to a state park. It includes the following variables:

- livebait: whether or not the group used live bait $(0/1)$,
- camper: whether or not the group brought a camper on their visit $(0/1)$,
- persons: the number of people in the group,
- child: the number of children in the group, and
- count: number of fish caught.

There are 250 observations in the data:

```
library(stats4nr)
fishing
```

```
## # A tibble: 250 x 6
##      nofish livebait camper persons child count
##       <dbl>    <dbl>  <dbl>   <dbl> <dbl> <dbl>
## 1         1        0      0       1     0     0
## 2         0        1      1       1     0     0
## 3         0        1      0       1     0     0
## 4         0        1      1       2     1     0
## 5         0        1      0       1     0     1
## 6         0        1      1       4     2     0
## 7         0        1      0       3     1     0
## 8         0        1      0       4     3     0
## 9         1        0      1       3     2     0
## 10        0        1      1       1     0     1
## # ... with 240 more rows
```

Histograms work well for plotting count data such as the number of fish caught. We can see that most campers catch zero or only a few fish and only a few groups catch many fish, e.g., greater than 20:

```
ggplot(fishing, aes(x = count)) +
  geom_histogram() +
  labs (x = "Number of fish caught",
        y = "Number of occurences")
```

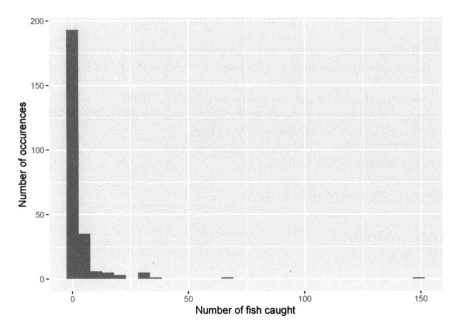

It is useful to quantify the mean and variance of the response variable to understand if dispersion is present:

```
mean(fishing$count)
```

```
## [1] 3.296
```

```
var(fishing$count)
```

```
## [1] 135.3739
```

In the case of the number of fish caught, we can say that **overdispersion** is present because the variance (135.4) is larger than the mean (3.3). **Underdispersion** would be present if the variance were less than the mean.

We are interested in predicting the number of fish caught using three predictor variables: the number of total people in the group, the number of children in the group, and whether or not the group used a camper during their visit. One might hypothesize that more people will increase the chances of fishing success, but more children, who are likely inexperienced and may not fish for long periods of time, may decrease the chances of fishing success. Similarly, groups that bring a camper for their visit may plan to camp for a longer duration, equating to more fishing time.

In R, the `glm()` function can be used to fit Poisson regression models to count data. The `family = "poisson"` statement within the function specifies a log link function in the model:

```
m.poisson <- glm(count ~ persons + child + camper,
                 family = "poisson", data = fishing)
summary(m.poisson)
```

```
##
## Call:
## glm(formula = count ~ persons + child + camper, family = "poisson",
##     data = fishing)
##
## Deviance Residuals:
##     Min       1Q   Median       3Q      Max
## -6.8096  -1.4431  -0.9060  -0.0406  16.1417
##
## Coefficients:
##               Estimate Std. Error z value Pr(>|z|)
## (Intercept)   -1.98183    0.15226  -13.02   <2e-16 ***
## persons        1.09126    0.03926   27.80   <2e-16 ***
## child         -1.68996    0.08099  -20.87   <2e-16 ***
## camper         0.93094    0.08909   10.45   <2e-16 ***
## ---
## Signif. codes:  0 '***' 0.001 '**' 0.01 '*' 0.05 '.' 0.1 ' ' 1
##
## (Dispersion parameter for poisson family taken to be 1)
##
##     Null deviance: 2958.4  on 249  degrees of freedom
## Residual deviance: 1337.1  on 246  degrees of freedom
## AIC: 1682.1
##
## Number of Fisher Scoring iterations: 6
```

The scale and magnitude of the coefficients confirm our hypotheses about the predictor variables and their relationship with fishing success. In addition to the output from the coefficients, also of note is the AIC value, a useful metric that can be compared to models fit with other distributions.

For fitting negative binomial regressions in R, the `glm.nb()` function available in the **MASS** package (Venables and Ripley 2002) is flexible. We can fit

a similar model form as the Poisson, but instead relying on the dispersion
parameter inherent to the negative binomial distribution:

```
library(MASS)

m.nb<- glm.nb(count ~ persons + child + camper,
              data = fishing)
summary(m.nb)
```

```
##
## Call:
## glm.nb(formula = count ~ persons + child + camper, data = fishing,
##     init.theta = 0.4635287626, link = log)
##
## Deviance Residuals:
##     Min       1Q   Median       3Q      Max
## -1.6673  -0.9599  -0.6590  -0.0319   4.9433
##
## Coefficients:
##             Estimate Std. Error z value Pr(>|z|)
## (Intercept)  -1.6250     0.3304  -4.918 8.74e-07 ***
## persons       1.0608     0.1144   9.273  < 2e-16 ***
## child        -1.7805     0.1850  -9.623  < 2e-16 ***
## camper        0.6211     0.2348   2.645  0.00816 **
## ---
## Signif. codes:  0 '***' 0.001 '**' 0.01 '*' 0.05 '.' 0.1 ' ' 1
##
## (Dispersion parameter for Negative Binomial(0.4635) family taken to be 1)
##
##     Null deviance: 394.25  on 249  degrees of freedom
## Residual deviance: 210.65  on 246  degrees of freedom
## AIC: 820.44
##
## Number of Fisher Scoring iterations: 1
##
##
##               Theta:  0.4635
##           Std. Err.:  0.0712
##
##  2 x log-likelihood:  -810.4440
```

Note that the scale and magnitude of the coefficients between the Poisson and NB models are similar, but with slightly different values. The dispersion parameter Theta is 0.4635 with a standard error of 0.0712.

We can also plot the predictions of number of fish caught from the Poisson and negative binomial models by applying the m.poisson and m.nb models to the **new.fish** data set. Specifying type = "response" provides the predicted number of fish caught:

```r
new.fish <- tibble(
  persons = c(1, 2, 3, 4, 1, 2, 3, 4),
  child = c(0, 0, 0, 0, 0, 0, 0, 0),
  camper = c(0, 0, 0, 0, 1, 1, 1, 1))



new.fish <- new.fish %>%
  mutate(count_p_pred = predict(m.poisson,
                                newdata = new.fish,
                                type = "response"),
         count_nb_pred = predict(m.nb,
                                 newdata = new.fish,
                                 type = "response"))
```

We can see that as the number of people in a group increases, so too does the number of fish caught. The Poisson model predicts a greater number of fish caught for groups that visit state parks with a camper, while for groups without a camper the negative binomial model predicts a greater number of fish caught:

13.2.2 Exercises

13.1 You hypothesize that as forests become older, their canopies become dominated by only a few species, which limits light availability for other species growing in the understory. Data to examine this hypothesis were acquired from the Hubachek Wilderness Research Center, an experimental forest located in northern Minnesota (Gill et al. 2019). Data were collected from 36 forest inventory plots with at least one tree larger than 12.7 cm in diameter. The number of species (num_spp), number of standing dead trees (num_dead), and average height of the trees in the plot (ht_m; meters) are the variables of interest.

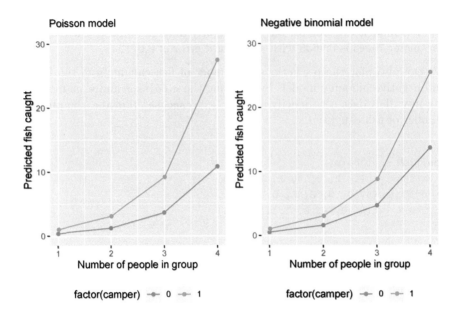

FIGURE 13.2 Model predictions of number of fish caught from the fishing data set.)

a. Run the code below to create the data set **sp__ht** to use in this analysis. Does the count variable of interest (num_spp) display characteristics of underdispersion or overdispersion? Comment on the usefulness of using the Poisson distribution for this variable.

```
sp_ht <- tibble(
  num_spp = c(4, 3, 2, 4, 2, 4, 2, 3, 2, 2,
              5, 1, 7, 4, 4, 7, 4, 6, 4, 5,
              3, 5, 5, 3, 7, 2, 5, 5, 3, 5,
              3, 5, 2, 4, 5, 4),
  num_dead= c(1, 1, 0, 0, 1, 0, 1, 1, 0, 0,
              2, 0, 0, 3, 0, 0, 0, 0, 0, 0,
              0, 0, 0, 0, 1, 0, 1, 0, 0, 0,
              0, 2, 0, 0, 0, 0),
  ht_m = c(17.6, 19.8, 26.0, 17.4, 19.8, 20.7,
           14.7, 17.6, 11.6, 19.4, 12.7, 16.4,
           12.6, 17.9, 23.2, 14.5, 18.6, 15.0,
           11.5, 15.1, 13.8, 9.9, 20.2, 7.3,
           6.2, 11.4, 16.5, 8.7, 9.8, 6.9,
```

```
          18.0, 10.0, 12.8, 13.7, 12.8, 16.7)
 )
```

b. Create a histogram for the `num_spp` variable and comment on its distribution.

c. Fit two count regression models predicting the number of species in a plot based on its average height. One model should assume the Poisson and the other a negative model distribution. Based on their AIC values, which model would you prefer and why?

13.2 Use the Poisson regression model you created in question 13.1 to make a prediction of the number of species for five additional forest inventory plots with mean heights of 12, 13, 19, 21, and 24 meters.

13.3 Make a graph using `ggplot()` that fits the Poisson model you created in question 13.1. For a range of tree heights from 5 to 30 meters along the *x*-axis, plot the predictions of the number of species for the model on the *y*-axis.

13.3 Zero-inflated models

To account for data with a high proportion of zeros, zero-modified count models, or **zero-inflated models** are often applied. For example, you may visit a forest to count the number of warblers you observe, but on repeated visits you observe none. Zero inflation is also apparent in the fishing data set, as the histogram indicates that zero is the most common value (i.e., the mode) for number of fish caught. These data result in excess zeros which make modeling with standard regression techniques a challenge.

In addition to visualizing them, calculating the percent of observations with zero as the response variable can help inform whether or not zero-inflated models should be considered. There are no guidelines or statistical tests that evaluate the specific number of observations required to perform zero-inflated regression. However, a general rule of thumb with natural resources data is that if 25% or more of your observations of a response variable are zero, zero-inflated models should be evaluated.

If the number of different response variable counts are relatively small (i.e., less than 30), summarizing the percent of observation by each count response can be done with the `summarize()` function. For the fishing data set, we calculate that 56.8% of the fishing groups caught zero fish:

```
fishing %>%
  group_by(count) %>%
  summarize(num_fish = n()) %>%
  mutate(Pct = num_fish / sum(num_fish) * 100)
```

```
## # A tibble: 25 x 3
##      count num_fish   Pct
##      <dbl>    <int> <dbl>
## 1        0      142  56.8
## 2        1       31  12.4
## 3        2       20  8
## 4        3       12  4.8
## 5        4        6  2.4
## 6        5       10  4
## 7        6        4  1.6
## 8        7        3  1.2
## 9        8        2  0.8
## 10       9        2  0.8
## # ... with 15 more rows
```

Zero-inflated models estimate two equations: a count model and a model describing the excess zeros. From these kinds of models, **zero-inflated Poisson (ZIP)** and **zero-inflated negative binomial (ZINB)** models are two types of models that estimate structural and stochastic zeros (Welsh et al. 1996; Gray 2005). In the ZIP model, zero counts are estimated by those derived from a binomial or Poisson distribution. Thus, the ZIP model has two parts: a Poisson count model and a logit model for predicting the excess zeros. A ZIP probability is estimated by:

$$f_{ZIP}(x) = \begin{cases} \pi + (1-\pi)e^{-\lambda}, x = 0 \\ (1-\pi)f_P(x), x = 1, 2, 3, ... \end{cases}$$

where π is the probability of zero occurrence, λ is estimated using independent variables, and $f_P(y)$ is the Poisson distribution.

Similar to the NB distribution, a ZINB distribution includes a dispersion parameter α. A ZINB probability is estimated by the mass function:

$$f_{ZINB}(x) = \begin{cases} \pi + (1-\pi)(\frac{1}{1+\lambda\alpha})^{1/\alpha}, x = 0 \\ (1-\pi)f_{NB}(x), x = 1, 2, 3, ... \end{cases}$$

where α is the dispersion parameter and $f_{NB}(y)$ is the negative binomial distribution.

Rather than modeling the excess zeros directly, some modelers will transform the data instead. After transforming a variable of interest, you could run a general linear model, however, the response variable will change. This will not necessarily improve linearity or assist the model meeting the assumption of homogeneity of variance. Furthermore, transformation of data with excess zeros cannot be accomplished with popular methods such as the logarithmic or reciprocal transformation.

13.3.1 Fitting zero-inflated models in R

The **pscl** package, abbreviated for the Political Science Computational Laboratory at Stanford University, fits a suite of models used in political science applications, including count models. The package is widely used to fit zero-inflated models. We can install and load the package to use it:

```
install.packages("pscl")
library(pscl)
```

The zeroinfl() function fits ZIP and ZINB models using syntax that is similar to the glm() function. By default, the zeroinfl() function will assume you wish to fit a ZIP model, which we can do with the **fishing** data:

```
m.zip <- zeroinfl(count ~ persons + child + camper,
                  data = fishing)
summary(m.zip)
```

```
##
## Call:
## zeroinfl(formula = count ~ persons + child + camper, data = fishing)
##
## Pearson residuals:
##      Min       1Q   Median       3Q      Max
## -3.05440 -0.74336 -0.44275 -0.07559 27.99304
##
## Count model coefficients (poisson with log link):
##             Estimate Std. Error z value Pr(>|z|)
## (Intercept) -0.79826    0.17081  -4.673 2.96e-06 ***
## persons      0.82904    0.04395  18.862  < 2e-16 ***
## child       -1.13666    0.09299 -12.224  < 2e-16 ***
## camper       0.72425    0.09314   7.776 7.51e-15 ***
##
```

```
## Zero-inflation model coefficients (binomial with logit link):
##              Estimate Std. Error z value Pr(>|z|)
## (Intercept)   1.6636     0.5155   3.227  0.00125 **
## persons      -0.9228     0.1992  -4.632 3.62e-06 ***
## child         1.9046     0.3261   5.840 5.21e-09 ***
## camper       -0.8336     0.3527  -2.364  0.01808 *
## ---
## Signif. codes:  0 '***' 0.001 '**' 0.01 '*' 0.05 '.' 0.1 ' ' 1
##
## Number of iterations in BFGS optimization: 12
## Log-likelihood: -752.7 on 8 Df
```

The primary item of note here is that two sets of coefficients are provided in the output: the first for the count model and the second for the zero inflation model. The count model coefficients for the ZIP model would be analogous to the coefficients found in the standard Poisson model (m.poisson) above.

The zero-inflated model coefficients are similar to a logistic regression as they are fit with the logit link function and address the question of whether the group caught zero fish or not. The signs of the coefficients for the zero-inflated model are opposite from the count model. This makes intuitive sense. For example, the persons coefficient in the count model (0.82904) indicates more people in a fishing group would catch more fish and the persons coefficient in the zero-inflated model (-0.9228) indicates more people in a fishing group would lower the probability of catching zero fish.

A ZINB model can be fit by setting the distribution to "negbin" in the zeroinfl() function:

```
m.zinb <- zeroinfl(count ~ persons + child + camper,
                   dist = "negbin", data = fishing)
summary(m.zinb)
```

```
##
## Call:
## zeroinfl(formula = count ~ persons + child + camper, data = fishing,
##     dist = "negbin")
##
## Pearson residuals:
##      Min       1Q   Median       3Q      Max
## -0.71806 -0.56103 -0.38173  0.04398 16.16365
##
## Count model coefficients (negbin with log link):
##              Estimate Std. Error z value Pr(>|z|)
## (Intercept)  -1.6178     0.3202  -5.052 4.36e-07 ***
```

```
## persons          1.0901      0.1117   9.761   < 2e-16 ***
## child           -1.2613      0.2473  -5.100 3.40e-07 ***
## camper           0.3857      0.2461   1.567 0.117108
## Log(theta)      -0.5929      0.1580  -3.753 0.000174 ***
##
## Zero-inflation model coefficients (binomial with logit link):
##              Estimate Std. Error z value Pr(>|z|)
## (Intercept) -12.0956    67.7886  -0.178    0.858
## persons       0.2906     0.7315   0.397    0.691
## child        11.0538    67.7088   0.163    0.870
## camper      -10.8729    67.7236  -0.161    0.872
## ---
## Signif. codes:  0 '***' 0.001 '**' 0.01 '*' 0.05 '.' 0.1 ' ' 1
##
## Theta = 0.5527
## Number of iterations in BFGS optimization: 44
## Log-likelihood: -395.5 on 9 Df
```

Note the similarity and differences between the ZIP and ZINB coefficients in the R output. The primary addition to the ZINB output is the dispersion parameter Theta = 0.5527. When implementing the zero-inflation model coefficients to make predictions on new data, a threshold or cutpoint probability needs to be specified to determine if a zero or non-zero prediction results. For this a standard 0.50 or 0.75 probability is often specified.

13.3.2 Exercises

13.4 The choice of which count regression model to implement depends in part on the characteristics of the response variable. Write R code and describe whether the following variables represent overdispersion, underdispersion, or equal dispersion. Also comment on which modeling technique may be appropriate for the data (e.g., Poisson, negative binomial, zero-inflated Poisson, or zero-inflated negative binomial).

  a. Ant species richness (spprich) in bogs and forests in New England, USA (available in the **ants** data set from the **stats4nr** package).
  b. The number of birds (NumBirds) observed in west-central California suburbs (available in the **birds** data set from the **stats4nr** package).
  c. The number of tree species (num_spp) observed in forest inventory plots in an experimental forest in Minnesota (available from question 13.1).

FIGURE 13.3 An American marten. Image: Bailey Parsons, Wikimedia Commons.

    d. The number of standing dead trees (num_dead) observed in forest inventory plots in an experimental forest in Minnesota (available from question 13.1).

13.5 Compare the Akaike's information criterion (AIC) for the zero-inflated Poisson and zero-inflated negative binomial models fit to the **fishing** data set. Which model is of better quality and why do you suspect the degrees of freedom differ between the two models?

13.6 A large number of standing dead trees in a forest is an indicator of quality habitat to sustain populations of American marten (*Martes americana*), a mammal species found in northern temperate and boreal forests. Using the data set provided from the experimental forest in question 13.1, answer the following questions.

    a. Write R code to determine the percentage of forest inventory plots that contain no observations of standing dead trees.

    b. Fit a model that predicts the number of standing dead trees (num_dead) in a forest plot using the average height of trees in a stand (ht_m). Fit a standard Poisson and zero-inflated Poisson model and compare them with AIC. Which model would you prefer?

13.4   Evaluating count regression models

While AIC is a useful metric to compare multiple count regression models (e.g., the Poisson, NB, ZIP, and NINB models), a number of other useful measures can determine the uncertainty and goodness of fit for count models. Diagnostic plots that display differences between predicted and observed counts are often employed to examine model fit of count regression models. Diagnostic plots can visualize the **deviance residuals**, showing trends in overestimations or underestimations across a range of values. Similarly, **Pearson's residuals** rely on the chi-square (χ^2) distribution and can be used to assess model fit. As an example, we can visualize the residual values from our model fit with a Poisson distribution and store them in the **fishing** data set. We first predict the number of fish and then determine the deviance and Pearson's residuals by changing the type = statement within the resid() function:

```
fishing <- fishing %>%
  mutate(count_pred = predict(m.poisson,
                            newdata = fishing,
                            type = "response"),
         res_dev = resid(m.poisson, type = "deviance"),
         res_pear = resid(m.poisson, type = "pearson"))
```

Then, we can plot the two residual values to determine trends and assess model fit:

```
ggplot(fishing, aes(count_pred, res_dev)) +
  geom_point() +
  labs(x = "Fitted value",
       y = "Deviance residual")
```

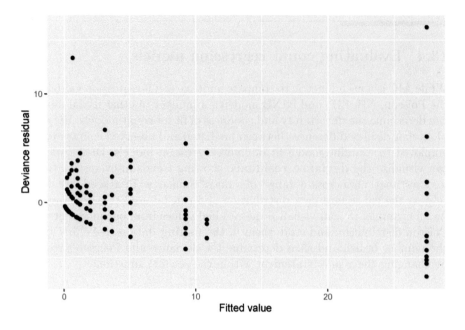

```
ggplot(fishing, aes(count_pred, res_pear)) +
  geom_point() +
  labs(x = "Fitted value",
       y = "Pearson residual")
```

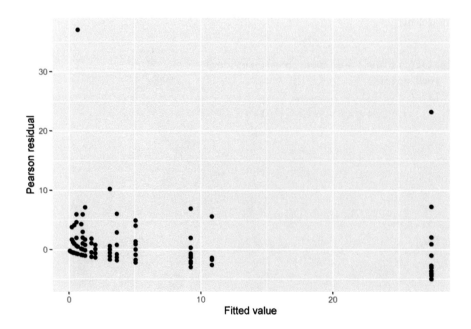

There is some concern with a single data point with a small estimated mean with a large residual (e.g., Pearson residual greater than 30 fish caught). The characteristics of this data point are worth looking into with more detail.

The root mean square error (RMSE) and mean bias (MB) are two metrics that allow you to determine the uncertainty of model predictions:

$$RMSE = \sqrt{\sum_{i=1}^{n} (y_i - \hat{y}_i)^2 / n}$$

$$MB = \sum_{i=1}^{n} (y_i - \hat{y}_i)/n$$

where y_i is the observed count, $\hat{y}_i$ is the estimated count, and n is the number of observations.

Pearson's chi-square (χ^2) statistic is a commonly used metric used to determine goodness of fit with count regression models.

$$\chi^2 = \sum_{i=1}^{n} \frac{(y_i - E(y_i))^2}{Var(y_i)}$$

The expected value $E(y_i)$ and variance $Var(y_i)$ can be calculated for the response variable of interest depending on the count model (e.g., Poisson, NB, ZIP, or ZINB; Zuur et al. [2009]). Using the (χ^2) statistic, data can be divided into intervals and the numbers of observations within each interval can be compared. Generally, a satisfactory model fit is indicated if the ratio of the χ^2 to its degrees of freedom is close to one.

13.4.1 Exercises

13.7 Create a diagnostic plot using `ggplot()` that shows the Pearson residuals for the zero-inflated Poisson model that predicts the number of fish caught from the **fishing** data set.

13.8 Write two functions in R that can be applied to any data set with observations and model predictions. One function should calculate the root mean square error (RMSE) and the other the mean bias (MB). Use the functions to determine the RMSE and MB for the standard Poisson and zero-inflated Poisson models that predict the number of fish caught from the **fishing** data set.

13.5   Summary

Count data, where observations represent non-negative integers, are widely found across natural resources disciplines. The Poisson distribution is the classic count distribution that can describe many distributions of count data. While it assumes that a mean value is equal to its variance, if this assumption cannot hold, consider if the data are distributed as a negative binomial distribution. When modeling data with a negative binomial, a dispersion parameter is specified to account for differences in values for the mean and variance. If a number of excess zeros are present in the data, consider the zero-inflated implementations of these models. Zero-inflated models estimate two equations: one that reflects the count model (analogous to the standard Poisson or negative binomial models) and another that estimates a logistic model for predicting the excess zeros.

Count models are generalized linear models and can be fit in R using code and syntax similar to logistic regression. The glm() function can fit standard Poisson models while the glm.nb() function from the **MASS** package fits negative binomial models. The **pscl** package is popular for fitting zero-inflated versions of these models and provides parsimonious output for interpreting and assessing models. It is a good practice to fit different versions of count models, including standard and zero-inflated ones, and compare their performance with metrics such as AIC and root mean square error and visualizations such as diagnostic plots. Having a thorough understanding of these regression modeling techniques will aid you when you encounter count data.

13.6   References

Gill, K.G., L.B. Johnson, R.A. Olesiak, R.J. Prange, and A.J. David. 2019. Defining the University of Minnesota experimental forests landbase. Retrieved from the University of Minnesota Digital Conservancy. Available at: https://hdl.handle.net/11299/209141.

Gray, B.R. 2005. Selecting a distributional assumption for modelling relative densities of benthic macroinvertebrates. *Ecological Modelling* 185: 1—12.

Venables, W. N., and B.D. Ripley. 2002. *Modern applied statistics with S. Fourth edition.* Springer, New York.

Welsh, A.H., R.B. Cunningham, C.F. Donnelly, and D.B. Lindenmayer. 1996.

Modelling the abundance of rare species: statistical models for counts with extra zeros. *Ecological Modelling* 88: 297—308.

Winkelmann, R. 2008. *Econometric analysis of count data.* Springer, Berlin.

Zuur, A., E.N. Ieno, N. Walker, A.A. Saveliev, G.M. Smith. 2009. *Mixed effects models and extensions in ecology with R.* Springer-Verlag, New York.

14

Linear mixed models

14.1 Introduction

Mixed models have emerged as one of the go-to regression techniques in natural resource applications. This is attributed to several reasons and primarily relates to the nested structure of natural resources data, either in a spatial or temporal context. First, technicians may collect information on experimental units that are close to one another in space or time. Second, oftentimes the same experimental units (e.g., the identical plant of animal) are measured again following their initial measurement. Third, mixed models can account for hierarchy within data. As an example, forest plots are often collected within stands, stands are located within ownerships, and a collection of ownerships comprise a landscape. Finally, mixed models allow the inclusion of both fixed and random effects. **Fixed effects** can be considered population-averaged values and are similar to the parameters found in "traditional" regression techniques like ordinary least squares. **Random effects** can be determined for each parameter, typically for each hierarchical level in a data set.

This chapter will discuss the theory and application of linear mixed models. Importantly, we'll learn how mixed models differ from ordinary least squares and other regression techniques, how to fit mixed models in R, and how to make predictions using mixed models in new applications.

14.2 Comparing ordinary least squares and mixed effects models

Recall back to simple linear regression where we used the concepts of least squares to minimize the residual sum of squares. This resulted in values for β_0 and β_1 that we considered constant or "fixed" (with some standard error associated with them).

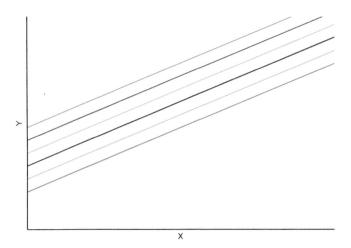

FIGURE 14.1 Example mixed model with random intercepts but identical slopes.

$$Y = \beta_0 + \beta_1 X + \varepsilon$$

The intercept (β_0) and slope (β_1) in simple linear regression will be chosen so that the residual sums of squares is as small as possible. For the same model in a mixed modeling framework, β_0 and β_1 are considered fixed effects (also known as the population-averaged values) and b_i is a random effect for subject i. The random effect can be thought of as each subject's deviation from the fixed intercept parameter. The key assumption about b_i is that it is independent, identically and normally distributed with a mean of zero and associated variance. For example, if we let the intercept be a random effect, it takes the form:

$$Y = \beta_0 + b_i + \beta_1 X + \varepsilon$$

In this model, predictions would vary depending on each subject's random intercept term, but slopes would be the same:

In another case, we can let the slope be a random effect, taking the form:

$$Y = \beta_0 + (\beta_1 + b_i)X + \varepsilon$$

In this model, the b_i is a random effect for subject i applied to the slope. Predictions would vary depending on each subject's random slope term, but the intercept would be the same:

One could also specify a random effect term on *both* the intercept and slope. This model form would be:

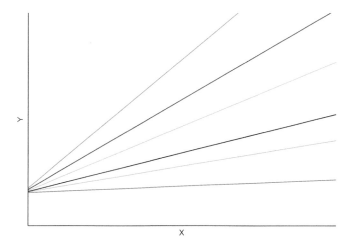

FIGURE 14.2 Example mixed model with random slopes but identical intercepts.

$$Y = (\beta_0 + a_i) + (\beta_1 + b_i)X + \epsilon)$$

In this model, a_i and b_i are random effects for subject i applied to the intercept and slope, respectively. Predictions would vary depending on each subject's slope and intercept terms:

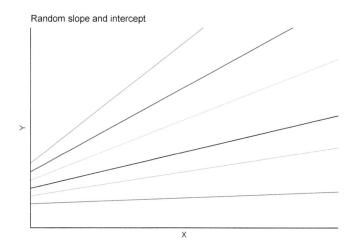

Random slope and intercept

FIGURE 14.3 Example mixed model with random intercept and slopes terms.

One of the most popular applications of mixed models is to account for the nested or hierarchical structure of data. As an example, measurement plots

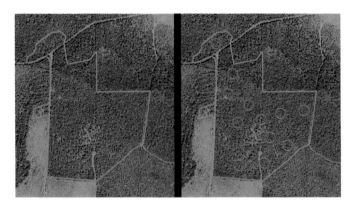

FIGURE 14.4 Forest stand (left) with measurement plots located throughout stand (right), showing nested design of data collection. Image: Ella Gray.

within a forest are identified in forest resource assessments and measured to
collect information of the status, size, and health of trees. These measurement
plots are nested within forest stands, or areas of trees a similar size and age.

Consider a random effect term applied to the intercept. By nesting subject j
within subject i (e.g., measurement plots nested withing stands), this model
form would be:

$$Y = (\beta_0 + b_i + b_{ij}) + \beta_1 X + \epsilon$$

In this model, b_i is the random effect for subject i and b_{ij} is the random effect
for subject j nested in subject i. In the forest plot example, we would obtain
a set of random effects for each forest stand i and a set of random effects for
each measurement plot nested within each stand ij. Hence, predictions would
result in two sets of random effects for each intercept (with identical slopes).
Consider an example application with three stands and four plots located
within those stands.

Even for a mixed model that uses one or only a few predictors, there are many
choices for specifying which parameters should be random. To quantify this,
you can fit several models with random effects applied to different parameters. After fitting models, you can evaluate them by assessing the quality of
each model using metrics such as the Akaike information criterion (AIC) or a
likelihood ratio test.

Another valid question is to ask whether mixed models are needed for your
analysis. For example, a simple linear regression model fit with ordinary least
squares may do. To examine this, fit a model with and without random effects.
The various model forms can be evaluated with AIC or a likelihood ratio test.
Lower AIC values for mixed model forms indicate that random effects are
useful in the predictions.

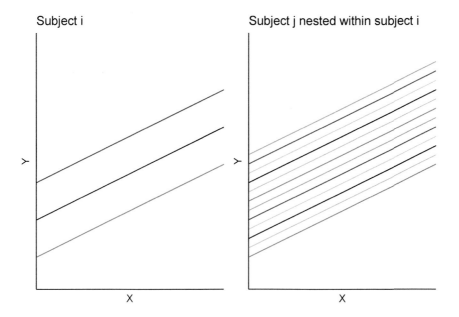

FIGURE 14.5 Example mixed model with random intercept terms applied to a nested design.

14.2.1 Exercises

14.1 Many applications of mixed models are compared to those fit with ordinary least squares. Using the **redpine** data set from the **stats4nr** package, we are interested in predicting a tree's height (HT) based on its diameter at breast height (DBH). Data are from 450 red pine (*Pinus resinosa*) observations made at the Cloquet Forestry Center in Cloquet, Minnesota in 2014 with DBH measured in inches and HT measured in feet.

    a. Write code using the ggplot() function that produces a scatter plot of red pine diameter and height. Add a linear regression line to the plot.

    b. Find the Pearson correlation coefficient between tree height (HT) and diameter at breast height (DBH).

    c. Fit a regression equation using ordinary least squares predicting tree height based on its diameter.

    d. Use the AIC() function to determine the Akaike information criterion of the fitted model.

FIGURE 14.6 A stand of red pine trees in northern Minnesota, USA. Image by the author.

14.3   Linear mixed models in R

Two popular R packages for performing mixed models in R are **lme4** (for linear models) and **nlme** (for linear and nonlinear models). To learn more about the theory behind mixed models, Pinheiro and Bates (2000) and Zuur et al. (2009) are excellent sources.

14.3.1   Case study: Predicting tree height with mixed models

We seek to build upon the analysis of question 14.1 by predicting the height of red pine trees based on their diameter. Data were collected from various forest cover types (CoverType) and forest inventory plots (PlotNum) across the forest: The **redpine** data set from the **stats4nr** package contain the data:

```
library(stats4nr)

head(redpine)
```

```
## # A tibble: 6 x 5
##   PlotNum CoverType TreeNum   DBH    HT
##     <dbl> <chr>       <dbl> <dbl> <dbl>
## 1       5 Red pine       46  13      51
## 2       5 Red pine       50   8.3    54
## 3       5 Red pine       54   8.2    48
```

```
## 4      5 Red pine     56  11.8    55
## 5      5 Red pine     63  9       54
## 6      5 Red pine     71  12.5    46
```

14.3.1.1   Random effects on intercept

We can expand a simple linear regression to a mixed model by incorporating the forest cover type from where a tree resides as a random effect. Cover types are assigned in the field based on the dominant species occurring in a plot, hence, we might have reason to believe that variability may be observed as one transitions from one cover type to the next. In the red pine data, the n_distinct() function indicates there are 13 different cover types on the property:

```
n_distinct(redpine$CoverType)
```

```
## [1] 13
```

We can see that while most of the red pine trees are found in the red pine cover type, the species is also found in the other 12 cover types, although less abundant:

```
ggplot(redpine, aes(DBH, HT)) +
  geom_point() +
  facet_wrap(~CoverType, ncol = 4) +
  labs(x = "Diameter at breast height (inches)",
       y = "Height (feet)")
```

The **lme4** package in R can be used to fit linear mixed models for fixed and random effects. We will use it to fit three linear mixed models that specify random effects on different parameters:

```
install.packages("lme4")
library(lme4)
```

The lmer() function is the mixed model equivalent of lm() and parameter estimates are fit using maximum likelihood. To specify the cover type as a random effect on the intercept, we write + (1 | CoverType) after specifying the independent variable DBH:

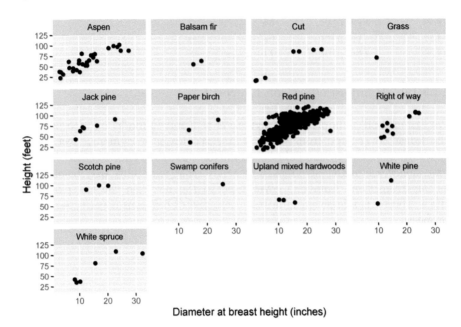

FIGURE 14.7 The relationship between tree height and diameter in 13 different cover types in Cloquet, MN.

```
redpine.lme <- lmer(HT ~ DBH + (1 | CoverType),
              data = redpine)
summary(redpine.lme)
```

```
## Linear mixed model fit by REML ['lmerMod']
## Formula: HT ~ DBH + (1 | CoverType)
##    Data: redpine
##
## REML criterion at convergence: 3652
##
## Scaled residuals:
##    Min     1Q  Median     3Q    Max
## -4.1165 -0.6923 -0.0591  0.6087  3.0488
##
## Random effects:
##  Groups   Name        Variance Std.Dev.
##  CoverType (Intercept)  52.24    7.228
##  Residual              191.14   13.825
## Number of obs: 450, groups:  CoverType, 13
##
```

```
## Fixed effects:
##              Estimate Std. Error t value
## (Intercept)   25.8867     3.2321   8.009
## DBH            3.0585     0.1169  26.169
##
## Correlation of Fixed Effects:
##      (Intr)
## DBH -0.537
```

The `lmer()` output contains similar-looking output compared to `lm()`. We can see from the output that the values for the fixed effects β_0 and β_1 are 25.8867 and 3.0585, respectively. The *Random effects* section contains details on the variance of the CoverType random effect and the residuals. Note the residual standard error of 13.825 feet, which can be compared to the value 14.31 obtained through fitting a model with ordinary least squares for question 14.1.

The `ranef()` function extracts the random effect terms from a mixed model. In this model, we can obtain the 13 random effects for each cover type:

```
ranef(redpine.lme)
```

```
## $CoverType
##                          (Intercept)
## Aspen                     -2.4313290
## Balsam fir                -6.4569721
## Cut                       -5.3398733
## Grass                      3.2321996
## Jack pine                  1.1099090
## Paper birch               -7.0258594
## Red pine                   6.4705981
## Right of way               0.7999610
## Scotch pine                8.9566777
## Swamp conifers            -0.3374007
## Upland mixed hardwoods    -0.9416721
## White pine                 6.9294464
## White spruce              -4.9656851
##
## with conditional variances for "CoverType"
```

As a visual, the HT-DBH models with varying random intercepts will show a different regression line for each cover type. We will specify a `predict()` statement within `geom_line()` to visualize the predictions.

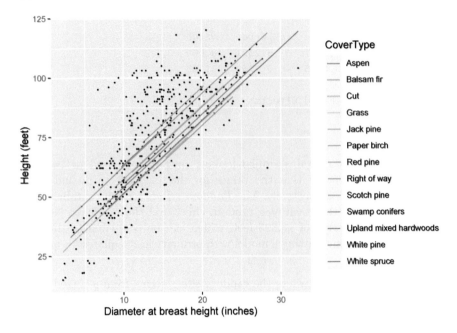

FIGURE 14.8 The height-diameter regression lines after fitting forest cover type as a random effect to predict red pine tree heights in Cloquet, MN.

```
ggplot(redpine, aes(DBH, HT)) +
    geom_point(size = 0.2) +
    geom_line(aes(y = predict(redpine.lme),
                  group = CoverType,
                  color = CoverType)) +
    labs(x = "Diameter at breast height (inches)",
         y = "Height (feet)")
```

Note that the lines have the same slope yet different intercepts, as depicted in the random effect term applied to β_0. You will notice that some predictions do not extend far through the x-axis, a reflection of the small sample size for red pine trees found in some cover types.

For the same tree diameter, the mixed model predictions indicate that trees growing in the Scotch pine (*Pinus sylvestris*) and red pine cover types would be tallest. Trees growing in the balsam fir and paper birch cover types would be shortest. Oftentimes with natural resources data, the distribution and magnitude of random effects can help to reinforce biological findings. For example, a wise forester may tell you that "pine trees grow well where pine trees should be growing". That is, taller red pine trees growing in pine cover types would be a likely outcome contrasted with red pine trees growing where fir and birch trees are better adapted.

14.3.1.2 Random effects on slope

An alternative mixed model could be specified by placing a random parameter on the slope term, which can potentially introduce model complexity. In the tree height example, we can specify (1 + DBH | CoverType) to place a random effect on the slope term for DBH:

```
redpine.lme2 <- lmer(HT ~ 1 + DBH + (1 + DBH | CoverType),
            data = redpine)
summary(redpine.lme2)
```

```
## Linear mixed model fit by REML ['lmerMod']
## Formula: HT ~ 1 + DBH + (1 + DBH | CoverType)
##    Data: redpine
##
## REML criterion at convergence: 3651.7
##
## Scaled residuals:
##     Min     1Q  Median     3Q     Max
## -4.0915 -0.6998 -0.0553  0.6024  3.0359
##
## Random effects:
##  Groups    Name         Variance  Std.Dev. Corr
##  CoverType (Intercept)   78.74861  8.8740
##            DBH            0.01029  0.1014  -1.00
##  Residual              190.92366 13.8175
## Number of obs: 450, groups:  CoverType, 13
##
## Fixed effects:
##              Estimate Std. Error t value
## (Intercept)  24.9363     3.7333   6.679
## DBH           3.1208     0.1235  25.269
##
## Correlation of Fixed Effects:
##     (Intr)
## DBH -0.706
## optimizer (nloptwrap) convergence code: 0 (OK)
## boundary (singular) fit: see ?isSingular
```

You can see that the error message boundary (singular) fit: see ?isSingular is shown, indicating that the model is likely overfitted. We may be trying to do too much by specifying the random effects on the slope. Hence, for these data, we might forgo the inclusion of a random slope parameter and instead focus on random effects for the intercept.

14.3.1.3 Nested random effects on intercept

We can expand the mixed model by incorporating the measurement plot (PlotNum) nested within forest cover type (CoverType) as random effects in the prediction of tree height. In the red pine data, there are 124 plots nested within the 13 cover types found in the data:

```
n_distinct(redpine$PlotNum)
```

```
## [1] 124
```

It would be challenging to view scatter plots that show the height-diameter relationship for each plot, however, adding facet_wrap() to ggplot code can visualize important differences in these trends. For example, note the distributions of red pine heights and diameters in plots 24 and 102. For the same range in diameters for both trees, note the taller heights for those in plot 24:

```
redpine %>%
    filter(PlotNum == c(24, 102)) %>%
    ggplot(aes(DBH, HT)) +
  geom_point() +
  facet_wrap(~PlotNum) +
  labs(title = "Red pine at Cloquet Forestry Center",
          subtitle = "HT-DBH by plot number",
          x = "Diameter at breast height (inches)",
          y = "Height (feet)")
```

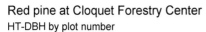

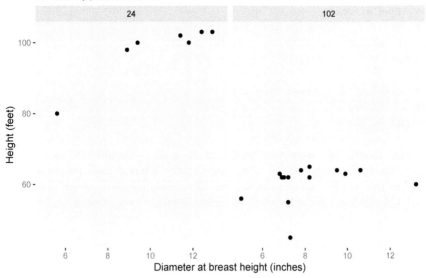

To specify nested random effects on the intercept (plot number nested within cover type), we can write (1 | CoverType/PlotNum) in the lmer() function:

```
redpine.lme3 <- lmer(HT ~ DBH + (1 | CoverType/PlotNum),
            data = redpine)
summary(redpine.lme3)


## Linear mixed model fit by REML ['lmerMod']
## Formula: HT ~ DBH + (1 | CoverType/PlotNum)
##     Data: redpine
##
## REML criterion at convergence: 3413
##
## Scaled residuals:
##     Min      1Q  Median      3Q     Max
## -3.8376 -0.4860  0.0388  0.5710  3.1585
##
## Random effects:
##  Groups            Name         Variance Std.Dev.
##  PlotNum:CoverType (Intercept) 123.63    11.119
##  CoverType         (Intercept)  19.17     4.378
##  Residual                       68.73     8.291
```

```
## Number of obs: 450, groups:  PlotNum:CoverType, 124; CoverType, 13
##
## Fixed effects:
##              Estimate Std. Error t value
## (Intercept)  30.5389     2.9080   10.50
## DBH           2.7111     0.1192   22.74
##
## Correlation of Fixed Effects:
##      (Intr)
## DBH -0.621
```

In this model, the estimated values of β_0 and β_1 are 30.5389 and 2.7111, respectively. Note that these values are slightly different from previous model fits. In this model, we can obtain the 13 random effects for each cover type and 124 random effects for each plot nested within each cover type.

```
ranef.lme3 <- ranef(redpine.lme3)
```

A quantile-quantile plot indicates the random effects are generally normally distributed:

```
plot(ranef.lme3)
```

```
## $`PlotNum:CoverType`
```

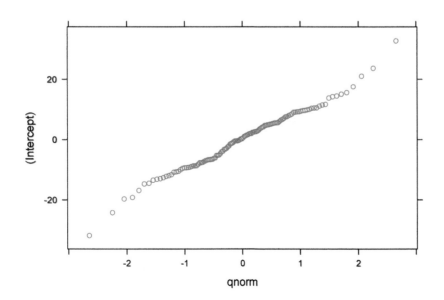

TABLE 14.1 Summary of ordinary least squares and linear mixed models fit with random effects applied to the intercept for the red pine data.

Model	Intercept	Slope	AIC
OLS	31.05	3.05	3676.1
Random = Cover Type	25.89	3.06	3660.0
Random = Cover Type/Plot	30.54	2.71	3423.0

```
##
## $CoverType
```

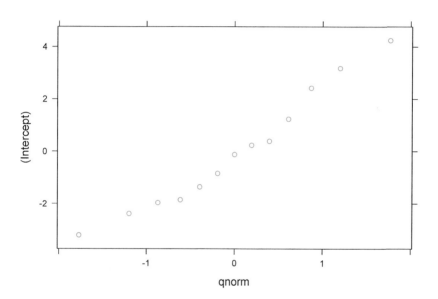

We can compare the AIC for each mixed model form fit with random effects applied to the intercept:

```
AIC(redpine.lme, redpine.lme3)
```

```
##               df       AIC
## redpine.lme    4 3659.952
## redpine.lme3   5 3422.994
```

Each model form, including the ordinary least squares model (from question 14.1) and linear mixed models is summarized in Table 14.1.

Note that the AIC is lowest for the mixed model with the plot number nested within cover type. While the emphasis here is on fitting linear mixed models, you may be interested in exploring the `glmer()` and `nlmer()` functions from **lme4** for fitting generalized linear and nonlinear mixed models, respectively.

14.3.2 Exercises

14.2 Use the **fish** data set from the **stats4nr** package to answer the following questions.

 a. Using `ggplot()`, make a scatter plot showing the weight of the fish (`Weight`; in grams) on the y-axis and the length from the nose to the beginning of the tail (`Length1`; in cm) on the x-axis. Color the observations for each of the seven species and add a `stat_smooth()` statement that displays a linear regression line for each species.
 b. Fit a linear mixed model predicting `Weight` based on `Length1` that incorporates `Species` as a random effect on the intercept.
 c. Fit a similar linear mixed model predicting `Weight` based on `Length1`, but this time apply the random effect of `Species` on the slope term.
 d. Compare the AIC values of the models in parts b and c. Which model may be preferred and why?

14.3 Use **Loblolly**, a built-in data set in R containing the height, age, and seed source of loblolly pine (*Pinus taeda*) trees, to answer the following questions.

 a. Fit a linear mixed model predicting tree height (`height`; in meters) based on tree age (`age`; years) that incorporates seed source as a random effect on the intercept. Create a quantile-quantile plot to assess whether the random effects are normally distributed.
 b. Ben Bolker created **broom.mixed**, an R package that extends what the **broom** package does to different kinds of model forms, including linear mixed models (Bolker 2021). Install and load the package, then use the `augment()` function to extract the fitted and residual values for the observations in **Loblolly**. Hint: consult section 8.6.2 where we do a similar activity with the **broom** package in an ordinary least squares regression.
 c. Create a scatter plot of residual and fitted values using `ggplot()` and add a smoothed trend line. What do you notice about the model and how might it be remedied?

14.4 The **lmerTest** package in R provides additional output and functions that work with linear mixed models fit with `lmer()`. After installing and loading the package, explore its `ranova()` function. This function performs a likelihood ratio test that assesses the contribution of random effects in a linear mixed model. Apply the function to the models fit with random effects on the intercept from questions 14.2 (the fish data) and 14.3 (the loblolly pine data). What is the result of the likelihood ratio tests, that is, do the random effects contribute to the performance of the model?

14.4 Predictions and local calibration of mixed models

14.4.1 Predictions for observations used in model fitting

Mixed models are advantageous because of their inclusion of random effects that may account for the hierarchy of data. If you are interested in making predictions for observations used in the model fitting, you have two options:

- **Use the fixed effects alone to make predictions.** This is identical to setting the random effect terms equal to zero.
- **Use the subject-specific random effects.** In this approach, you create best linear unbiased predictions (BLUPs) to estimate random effects for linear mixed models.

As an example, consider we wish to make two sets of predictions for red pine heights using the model with a random effect for cover type applied to the intercept. Note that the coef() function applied to this model reveals the different intercepts for the 13 cover types with an identical slope:

```
coef(redpine.lme)
```

```
## $CoverType
##                          (Intercept)     DBH
## Aspen                       23.45539 3.05847
## Balsam fir                  19.42975 3.05847
## Cut                         20.54684 3.05847
## Grass                       29.11892 3.05847
## Jack pine                   26.99663 3.05847
## Paper birch                 18.86086 3.05847
## Red pine                    32.35732 3.05847
## Right of way                26.68668 3.05847
## Scotch pine                 34.84340 3.05847
## Swamp conifers              25.54932 3.05847
## Upland mixed hardwoods      24.94505 3.05847
## White pine                  32.81616 3.05847
```

```
## White spruce          20.92103 3.05847
##
## attr(,"class")
## [1] "coef.mer"
```

We will make one set of predictions using the fixed effects only and another using the subject-specific random effects. In R, the predict() function makes model predictions from different model types. For predictions from models fit with the lmer() function, a minimum of three arguments are needed:

- First, specify the R object that stores the model (e.g., redpine.lme).
- Second, specify the data set in which to apply to the predictions. In our case, we will add a new column to the original **redpine** data set.
- Lastly, use the re.form statement to specify whether you want to make predictions using fixed effects only or predictions using subject-specific random effects (e.g., BLUPs). If re.form = NULL, predictions are made including all random effects. If re.form = NA, predictions are made using fixed effects only (equivalent to setting the random effects equal to zero).

We can apply these new predictions by creating new variables using mutate() in the **redpine** data set and printing the first 10 rows of data using the top_n() function:

```
redpine <- redpine %>%
   mutate(HT_pred_fixed_re = predict(redpine.lme,
                                     redpine,
                                     re.form = NULL),
          HT_pred_fixed = predict(redpine.lme,
                                  redpine,
                                  re.form = NA))

redpine %>%
   top_n(10)
```

```
## Selecting by HT_pred_fixed

## # A tibble: 10 x 7
##    PlotNum CoverType   TreeNum  DBH   HT HT_pred_fixed_re HT_pred_fixed
##      <dbl> <chr>         <dbl> <dbl> <dbl>           <dbl>         <dbl>
## 1       40 White spruce      4  32.3   104            120.          125.
## 2       41 Red pine          1  27.7   108            117.          111.
## 3       45 Red pine         13  27.3   102            116.          109.
## 4      208 Red pine          1  26.9   107            115.          108.
## 5      236 Red pine          7  26.1   112            112.          106.
```

```
## 6    275 Aspen        4  27.8   88           108.        111.
## 7    306 Red pine    20  26.2  101           112.        106.
## 8    316 Red pine    10  28.3   62           119.        112.
## 9    320 Red pine     2  27    114           115.        108.
## 10   340 Red pine     1  25.8   97           111.        105.
```

Note the differences in the predictions of heights based on the two approaches.

14.4.2 Predictions for observations not used in model fitting

For making predictions outside of the original data used in model fitting, there are several approaches. First, you can make predictions using fixed effects, only. This is the most common approach one would choose when applying a mixed model. As we'll learn shortly, mixed models can be locally calibrated, however, a subsample of the response variable is needed and is not often available. Oftentimes, with natural resources data, we implement experiments to monitor outcomes that are not immediately observed, hence, local calibration is not possible and we must rely on fixed effects predictions.

If a subset of measurements of the response variable are available, a second approach involves local calibration to improve the precision of estimates. With data that are derived from diverse populations, such as a tree of a single species from a forest with 100 plant species, local calibration can better reflect local environmental conditions and microclimates. For example, Temesgen et al. (2008) observed that a locally calibrated mixed model predicting the height of Douglas-fir trees outperformed predictions using fixed effects only.

Local calibration can be conducted after taking a subsample of the response variable (e.g., the height of the tree) and determining the estimated BLUPs for the mixed model. The drawback of this approach is that new measurements need to be obtained for the variable you seek to predict. However, researchers have observed that using even a single observation to locally calibrate a mixed model leads to better precision in estimates (Trincado et al. 2007, Temesgen et al. 2008). An excellent worked example predicting tree height using local calibrations for a nonlinear model form can be found in VanderSchaaf (2012).

A final approach, which may see limited applications depending on the variable chosen for the random effect, involves using the estimated random effects from the model fit in the prediction of new observations. In our analysis of red pine tree heights, we could use the estimated random effects for cover type to make new predictions for tree heights found in that cover type. For example, Kuehne et al. (2020) observed that using random effects from a mixed model predicting tree diameter growth with species as a random effect outperformed a fixed effects model fit only to that species. In addition, making growth predictions by extracting the random effects was robust to account for tree growth of species with only a few observations. Using the **lmer** package, this would be equivalent to using the `predict()` function on a data set of new observations and specifying `re.form = NULL`.

This final approach has a few drawbacks in that you may have observations in a new data set from subjects that were not contained in the data set used in model fitting. For example, you may encounter a new species which you have not estimated a random effect for. The choice to use this approach would also depend on the variable chosen for the random effect in the model. For example, you might assume that tree species and cover type can be identified in future observations you might make, however, oftentimes random effects are specified to account for attributes that are less biologically meaningful. For example, unique values such as the measurement plot, blocking unit, or replication unit are often specified as random effects in natural resources studies. It is impractical to use the extracted values from these random effects and apply them to new observations. If this approach is not practical, consider making predictions using fixed effects only or locally calibrating.

14.4.3   Exercises

14.5 Recall the linear mixed model you created in question 14.2c that predicted fish weight based on its length with the random effect of species on the slope term. Answer the following questions using the **fish** data set.

    a. Write code in R to predict `Weight` based on `Length1` using fixed effects, only.

    b. Write code in R to predict `Weight` based on `Length1` using the estimated BLUPs that incorporate `Species` as the random effect.

    c. Calculate the mean bias of each prediction of fish weights. Mean bias can be obtained by calculating the difference between each observation and its prediction and then averaging these values. Based on the calculations, on average which prediction approach would you prefer?

14.6 You have collected ten more observations on the ages of loblolly pine trees and wish to apply the linear mixed model you created in question 14.3 to predict their heights. The data were collected from three seed sources that were also measured in the data set used in model fitting:

    a. Write code in R to make two predictions of loblolly pine tree heights using this new data set: one using fixed effects only and the other using the extracted random effects from the model you fit that incorporated `Seed` as the random effect.

TABLE 14.2 Ten new observations of loblolly pine tree ages from different seed sources.

Seed	age
305	7
305	9
305	15
321	5
321	6
321	19
327	4
327	12
327	22
327	24

b. Make a scatter plot of predicted tree heights (y-axis) along the range of tree ages (x-axis) for these 10 observations. Add a `col =` statement to your code in `ggplot()` to make two colors for each of the different kinds of model predictions. Explain the differences you observe in the predictions in a sentence or two.

14.5 Opportunities and limitations of mixed models

We have discussed many of the benefits of mixed models, including their ability to account for data collected in a hierarchical structure and to provide more precise predictions compared to regression using ordinary least squares. These reasons alone make them popular analytical techniques to use with natural resources data. Other potential applications of mixed models include using them to predict missing values. For example, mixed models can be fit to a data set and then locally calibrated to estimate missing values. Mixed models are also fitting to analyze experimental data with blocking and replicate effects in addition to studies that make repeated measurements on the same subjects over time (West et al. 2015).

Despite their popularity, there are drawbacks to the use and implementation of mixed models. While computational resources have expanded greatly in recent years, fitting mixed models remains computationally demanding in some settings. In particular with natural resources data, models fit to data sets with a large number of observations (e.g., greater than 100,000) with multiple variables can often fail to converge or encounter problems during the model fitting process. Random effects applied to slope terms often require more time and effort to converge compared to when random effects are applied to intercept.

Mixed models also suffer from the need to continually evaluate them with other regression techniques. Mixed models often need to be evaluated with other model forms, including comparisons with models fit with ordinary least squares and those with and without random effects. Mixed models also may need to be refit iteratively to determine which parameters should be specified as random in a model. While additional effort may be needed to decide whether or not mixed models are appropriate for your analysis, functions that provide useful metrics such as AIC and likelihood ratio tests can aid in the determination of the applicability of mixed models in your own work.

14.5.1 Exercises

14.7 While much of this chapter has presented mixed models in the context of regression, they can also be implemented to analyze experimental treatments, similar to analysis of variance. To see this in practice, install the **MASS** package and load the **oats** data set found within it. These data contain the yield of oats from a field trial with three oat varieties and four levels of nitrogen (manure) treatment. Type print(oats) and ?oats in R and investigate the data. Create a linear mixed model using the lmer() function that predicts oat yield with variety, nitrogen treatment, and their interaction as fixed effects. Specify the experimental block as a random effect on the intercept. Use the anova() function to print the results. What roadblocks do you encounter when looking to assess the significance of the variety and nitrogen treatments?

14.8 The **nlme** package performs linear and nonlinear mixed effects modeling. Investigate the package by typing ?nlme and perform the same linear mixed model as in question 14.7 using its lme() function. At a level of significance of $\alpha = 0.05$, do the variety, nitrogen treatments, and their interaction influence oat yield?

14.9 The **CO2** data set records the carbon dioxide uptake in different plant grasses subject to varying treatments. Type print(CO2) and ?CO2 in R and investigate the data. Say you are interested in creating a mixed model to predict uptake based on all of the other variables in the data. How would you specify your model and which parameters/variables would you consider fixed and random effects?

14.6 Summary

Natural resources data are fitting to analyze with mixed models because data are often nested and subjects are often measured repeatedly though time. Mixed models differ from traditional regression techniques such as ordinary least squares by including random effects terms that can be applied to different parameters within a fitted equation. The **lme4** and **nlme** packages in R are some of the most popular ones for fitting and making inference with mixed models.

Making new predictions with mixed models presents several options for an analyst. If a subsample of the response variable is available, mixed models can be calibrated with local observations to result in a more precise estimate. More often, additional data is not practical to obtain, hence, predictions using mixed models can be made using only the fixed effects or population average values. The implementation and application of mixed models requires a thorough evaluation to determine whether or not random effects are needed, and if so, which parameters should be specified as random. The application of mixed models in natural resources can help to fill in missing data, analyze data from experimental trials, and account for the unique attributes of data collected in a longitudinal format.

14.7 References

Bolker, B. 2021. Introduction to **broom.mixed**. Available at: https://cran.r-project.org/web/packages/broom.mixed/vignettes/broom_mixed_intro.html

Kuehne, C., M.B. Russell, A.R. Weiskittel, and J.A. Kershaw. 2020. Comparing strategies for representing individual-tree secondary growth in mixed-species stands in the Acadian Forest region. *Forest Ecology and Management* 459: 117823.

Pinheiro, J.C., and D.M. Bates. 2000. *Mixed-effects models in S and S-PLUS*. Springer-Verlag, New York.

Temesgen, H., V.J. Monleon, and D.W. Hann. 2008. Analysis and comparison of nonlinear tree height prediction strategies for Douglas-fir forests. *Canadian Journal of Forest Research* 38: 553–565.

Trincado, G., C.L. VanderSchaaf, and H.E. Burkhart. 2007. Regional mixed-effects height-diameter models for loblolly pine (*Pinus taeda* L.) plantations. *European Journal of Forest Research* 126: 253–262.

VanderSchaaf, C.L. 2012. Mixed-effects height-diameter models for commercially and ecologically important conifers in Minnesota. *Northern Journal Applied Forestry* 29: 15–20.

West, B.T., K.B. Welch, and A.T. Galecki. 2015. *Linear mixed models: a practical guide using statistical software, 2nd ed.* Chapman and Hall/CRC, Boca Raton, FL. 440 pp.

Zuur, A., E.N. Ieno, N. Walker, A.A. Saveliev, and G.M. Smith. 2009. *Mixed effects models and extensions in ecology with R.* Springer-Verlag, New York. 574 pp.

15

Communicating statistical results with visualizations

15.1 Introduction

As we have discussed, good statistical analyses begin and end with visualizations. By visualizing your data after conducting statistical analyses, your audience will understand your findings, provided you have displayed the results in a clear and succinct format. Ending a statistical analysis with quality visualizations of the results typically represents the end of the data analysis life cycle:

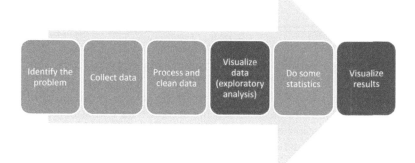

FIGURE 15.1 The data analysis life cycle.

Do not underestimate the importance and dedication needed to create and share quality statistical graphics:

"Anyone can run a regression or ANOVA! Regression and ANOVA are easy. Graphics are hard."

This quote by statistician Andrew Gelman (2011) is overwhelmingly true, particularly as we live in a time when data can be viewed, downloaded, and analyzed on a seemingly endless basis. Just as there are minimum standards for creating good graphics, such as choosing the appropriate type of figure to plot the data and labeling x- and y-axes appropriately with units, we can easily do "too much" with our graphics, e.g., by plotting too many colors in a single figure or not describing our graphics with enough information.

It is disappointing when quality statistical analyses are represented poorly with inadequate visualizations. Researchers and technicians likely spend considerable time and effort in collecting, summarizing, and analyzing natural resources data. It can be frustrating when all of that time and effort goes wasted when results are not presented clearly. This chapter will discuss some concepts and strategies for creating effective visualizations that resonate with your audience.

15.2　Strategies for presenting statistical results clearly

For all of the work and effort many of us go through in learning and applying quantitative methods to data, presenting statistics for statistics sake is rarely the primary motivation of our work. We use statistics to help inform other processes to better understand how the world works. For example, an analysis of variance might be performed to help understand the impacts of an experimental treatment on a response variable. Unfortunately, your audience likely does not care to know the specific p-values associated with your output, but they are certainly interested in evaluating the impacts of the experimental treatment and what it means to their costs, time, and effort.

There has recently been an increased recognition of the importance of storytelling in business and academic environments. Before, during, and after conducting your statistical analysis, be cognizant of the story that your analysis will help tell. Kurnoff and Lazarus (2021), authors of the book *Everyday Business Storytelling*, provide evidence which states that people remember stories, not data. Lean on your data and statistical output to tell a larger story about the process or phenomena under examination.

Consider your audience when presenting your statistical results. A colleague, collaborator, or supervisor may be interested in seeing your statistical output directly from R, but the majority of the people interested in your work prefer to see a summarized version of it that is presented in an appealing way. Assume that anyone that does not have a natural resources degree cannot understand the jargon used within your discipline. When creating and presenting graphs for the general public, use easy-to-understand units and supplement your visualizations with photos and text to convey key messages.

Data and visualizations have the benefit in that they can help persuade others and provide a research-based approach to inform decision making. But using too many visualizations can overwhelm your audience, lowering the effectiveness of your data and potentially losing the trust and confidence in you, the data analyst. This is often found in scientific and business presentations, with a term that Kurnoff and Lazarus (2021) term the "Frankendeck," a slide deck with a hodgepodge of information, including too much data and visualizations. Presentations and reports that do not distill and synthesize statistical results can be rendered incoherent and ineffective. Be economical with the amount of statistical results and visualizations you present, but always be willing to share more detailed results with those that ask.

15.2.1 Exercises

15.1 The same data set or summary of a statistical analysis can be used to create different visualizations depending on your audience. Using a data set or results from a statistical analysis that you are familiar with, create two to three different visualizations using tables or figures of different styles. For example, one visualization could show a bar plot displaying mean values and error bars denoting standard errors across different levels of a categorical variable. This visualization might be appropriate for "the researcher." Another visualization could present a word cloud of the data showing differences in the number of observations within each categorical variable. This visualization might be appropriate for a broader audience, i.e., "the general public."

15.2 Find a presentation slide deck you recently delivered that included data, or find one online to use as a case study. With respect to the data and statistical results presented in the slides, make several notes for how you can improve its layout and design. Share a specific slide on social media and ask your friends and followers for their candid feedback on how it can be improved.

15.3 Explore several popular websites and blogs dedicated to data visualization and presenting data effectively. Make notes about how you can improve your own data visualizations. A few worth exploring include:

- FlowingData: `https://flowingdata.com/`
- Evergreen Data: `https://stephanieevergreen.com/blog/`
- The Economist: `https://www.economist.com/graphic-detail`
- Storytelling with Data: `https://www.storytellingwithdata.com/blog`

15.3   Choosing between a table, figure, or data dashboard

Tables, figures, and data dashboards are three common methods to present data and statistical results. We can use them in different ways.

Tables are the preferred option under three scenarios. First, when we want our audience to know precise numbers or text, tables are best. As an example, when purchasing a brand new car, we would like to know that it costs exactly $22,495, not "about $20,000." Second, tables are useful to compare individual values, say values from one experiment compared to another. Lastly, tables are also useful when we are presenting several variables each with different units of measure. As an example, a table might include the number of visitors to a national park, how many days they're staying at the park, and how much money they spent during their stay.

When you construct tables, you should follow these guidelines, which are common across nearly all professions:

- Every table should have a **caption** that is numbered (typically placed at the top of the table).
- Tables should have **column headers** that are labeled with their appropriate units.
- The **cells** of a table should contain the values from the data or a summary of it.
- **Footnotes** should be used to explain what you want your audience to understand from a specific attribute in the table.

Figures are the preferred option under two scenarios. First, when trends or relationships between different values exist, figures are appropriate. For example, figures are fitting to show how values change over time, or how relationships exist between the same values for different populations. Second, figures are useful for photographs, maps, or other illustrations that might be a part of a project.

Figures have an advantage in that color can be added to enhance the interpretation of results. Follow these guidelines to construct effective figures:

- Similar to tables, every figure should have a **caption** that is numbered (typically placed at the bottom of the figure).
- Each figure should have a clearly labeled x **and** y **axis** with their appropriate units.
- If a **legend** appears on the figure, be sure it is labeled appropriately.

It is best to adhere to a few best practices when creating tables and figures. First, you want to make tables and figures stand alone. That is, the audience should be able to understand your message by looking only at the table or figure by itself. If someone navigates to a random page in your report, you want them to understand the key takeaways without having to read the entire report. Second, by numbering all of your tables and figures, you can refer to them in your text. Third, when using captions, do not assume your audience knows all of the abbreviations you use, so be sure to write out all of them. Fourth, there is no need to repeat information in one table or figure that is already in another table or figure. It is a good practice to be economical with your results and respectful of your audience's attention. Finally, keep the same general format and style throughout your written report and document. This includes using the same style of fonts, headings, and other attributes that are found in tables and figures.

Interactive data dashboards have arisen as a popular method to disseminate large amounts of information in a variety of formats. Whereas tables and figures are static representations of data, interactive data visualizations allow a user to select which variables and types of graphics they are interested in seeing. Interactive data visualizations are best used with data sets that contain a large number of observations and numerous variables.

To create data dashboards, the **shiny** package creates web applications using R code. It contains a user interface that can be customized depending on user needs. Visualizing your data with a dashboard can help you and others explore your data, evaluate different statistical models and their fit to validation data, and investigate statistical oddities such as outliers and unexpected distributions of data.

You can begin by creating simple web applications designed with the **shiny** package, then move on to more complex ones depending on you and your audience's needs. The best resource for learning the package is the text *Mastering Shiny* by Wickham (2022).

Data can also be analyzed in R and used in other applications that showcase data. StoryMaps developed by Esri are popular across many natural resources disciplines, in particular because ArcGIS software is widely used across resource management agencies and companies. One excellent example is the US Department of Agriculture Forest Service's (2022) *A guide to the forest products industry of the southern United States*, a StoryMap that displays timber product output data and the locations of primary wood processing plants and

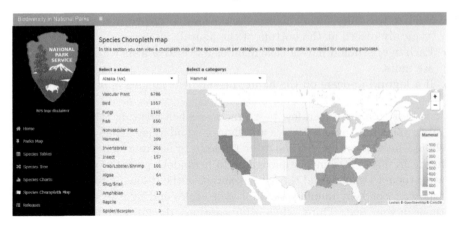

FIGURE 15.2 Example web application designed with the shiny package in R displaying the biodiversity across taxa in US National Parks. Application created by Alessio Benedetti.

mills across the region. The Texas A&M Forest Service (2022) has similarly used ArcGIS software to create *My City's Trees*, a web application that uses tree inventory data from urban areas and allows users to produce custom analyses and reports.

15.3.1   Evaluating statistical graphics

As you develop your own statistical graphics and observe visualizations developed by others, think critically about the elements that are effective in conveying the results. Wickham (2011) outlines the four C's of critiquing statistical graphics, as discussed in the next sections.

15.3.1.1   Content

With content, ask yourself: how do I want to present my data? As an example, here are two formats to present tree heights from the **elm** data set:

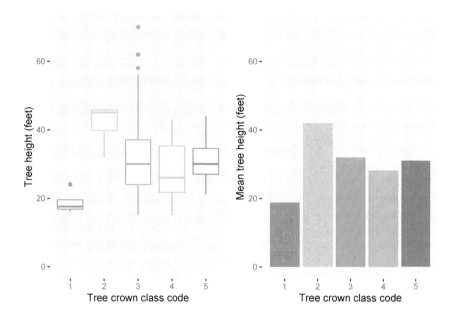

These figures present the tree height by crown class code indicating the relative crown position of the tree: open grown, dominant, co-dominant, intermediate, or suppressed. On the left, we see the distribution of observations in the form of box plots. We can see that there is a relationship between the two variables. On the right, we see those data summarized. This bar chart shows the mean tree height by crown class. When constructing your own tables and figure, consider whether the content should include a depiction of the raw data (i.e., the box plot) or a summarized version of it (i.e., the bar plot).

15.3.1.2 Construction

Fortunately, most of the default settings available in the **ggplot2** package will allow for the construction of effective figures. In some cases, adding different themes and elements to figures may be required. Become acquainted with the components of tables (e.g., captions, column headers, and footnotes) and figures (e.g., axis labels, captions, and legends) and how to change their attributes using the `ggplot()` function. With natural resources data, a few common elements of graphics that are often modified within the `ggplot()` function include:

- changing the position of the legend (e.g., `theme(legend.position = "top")`) or removing the legend entirely (`theme(legend.position = "none")`),
- changing the attributes of the x or y-axis (with `scale_x_continuous()` or `scale_y_continuous()`),

- printing a panel of figures with a single data set (using `facet_wrap()` or `facet_grid()`) or multiple data sets (using functions available in the **patchwork** package), and
- altering the color palette of your graphs to make them friendly to color-blind individuals (using functions available in the **colorspace** package).

For additional methods to customize your graphics with **ggplot2**, consult Wickham (2016).

15.3.1.3 Context

When you think about the context of a table or figure, consider whether it represents an idea, or if it represents real data that you collected. R software is quite good for plotting real data that exists in data sets, but consider other software to create visualizations that represent concepts or ideas. For example, flow charts and Venn diagrams can represent thought processes and concepts that do not necessarily rely on quantitative data.

15.3.1.4 Consumption

When you think about the consumption of a visualization that you design, ask yourself: who is the audience? Are there some audiences that prefer one kind of visualization over another? What is the format of the table or figure? Is it in a report, presentation, or manuscript? Always think about how the audience will be consuming your table or figure.

As an example, you don't want to include a table with 20 columns and 50 rows of data in a presentation on a slide that you show for 10 seconds. That's simply not enough time for the audience to digest that information. Instead, use a table in a presentation that only shows the key results, and perhaps put the large table with 50 rows of data in the appendix of a written report.

15.3.2 Exercises

15.4 The `kable()` function available in the **knitr** package in R creates simple tables for a number of documents. Learn more about the `kable()` function by reading through the section labeled *Tables* in Xie et al. 2019: `https://bookdown.org/yihui/rmarkdown/r-code.html#tables`. Use the **ant** data set from the **stats4nr** package to create the following tables:

  a. A table that displays the first eight rows of the **ant** data set.
  b. A table that displays the mean values for species richness in the bog and forest ecosystems, with an appropriate caption.

15.5 Find a figure from a scientific article or report that you recently read. Try to "copy the artist" by re-creating the figure from the article and coding it in `ggplot()`. Inspect the figure and note the approximate values contained in the figure. Choose a relatively simple figure, i.e., a bar plot or line graph where you can find the values for relatively few data points. If you require more precise values, web applications like *eleif* (https://eleif.net/photo_measure.html)[1] can be used to measure specific distances on an image.

15.6 Explore the gallery of Shiny applications available at RStudio's webpage: `https://shiny.rstudio.com/gallery/`. Which two or three Shiny projects resonate with you in how they allow a user to interact with the data?

15.4   Automate your work

Automation of your analytical work can be described as employing automatic processes and systems for collecting, analyzing, and archiving data. Although the two terms are not necessarily interchangeable, automation and reproducibility of your work often can be done in parallel. Your analysis is reproducible when others can reproduce the results if they have access to the original data, code, and documentation (Essawy et al. 2020).

15.4.1   Make your work reproducible

Integrating concepts of reproducibility in your work sends the signal that your analysis is high quality, trustworthy, and transparent (Sandve et al. 2013). In addition to being helpful to others, integrating actions to increase the reproducibility of your work also helps you as an analyst to better understand where you "left off" on an analysis as you spend time away from working on it. Some example actions you can take to improve your reproducibility are outlined in Powers and Hampton (2019). A few example ones include:

- Engage in sound data management practices with a plan in place on how you will collect, store, analyze, and archive data.
- Use coding scripts in R to "wrangle" data in their raw format, and annotate and comment within your code.
- Use version control systems for your R code such as Git (https://gitscm.com/)[2] and GitHub (https://github.com/)[3].
- Share code with collaborators and colleagues to promote reproducibility.

---

[1] https://eleif.net/photo_measure.html
[2] https://gitscm.com/
[3] https://github.com/

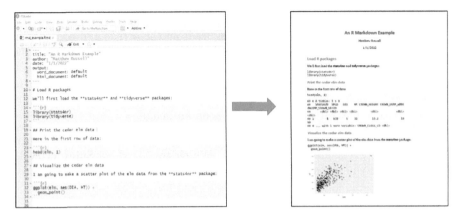

FIGURE 15.3 Example R Markdown code from RStudio (left) that can be knit into a document in Microsoft Word (right).

In R, one of the best tools that facilitates automation and reproducibility of your work is R Markdown (Xie et al. 2019). An R Markdown document is written with markdown, a coding style that uses plain text format. The value in using R Markdown is that you can embed R code, tables and figures, and written text to create a dynamic documents. These documents can include PDFs, Microsoft Word documents, HTML websites, and books (like this one). After code is written with R Markdown, it can be knit to one of the various output documents.

If you're using RStudio in your R session, the **rmarkdown** package comes preloaded. To start a new R Markdown document in RStudio, navigate to *File > New File > R Markdown* in the toolbar. Within these R Markdown files (.Rmd extension), headings and sections can be labeled with hashtags, text can be written, and R code can be written into "chunks."

Syntax in R Markdown files can be specified to make text appear in italics, bold, and other common settings to create written text. R code that performs statistical operations can be suppressed in a knitted document so that your audience does not see all of the code written to produce a report. For performing repetitive analyses such as running a new analysis on the same data structure (e.g., preparing a quarterly sales report), R Markdown documents are well suited to make these tasks reproducible, saving you time and effort in the future.

15.4.2   Automate your analysis workflow

Implement methods and techniques that automate your statistical analyses. Before your statistical analysis, write code to visualize its characteristics. Importantly, write this code as a series of functions which you can use in a

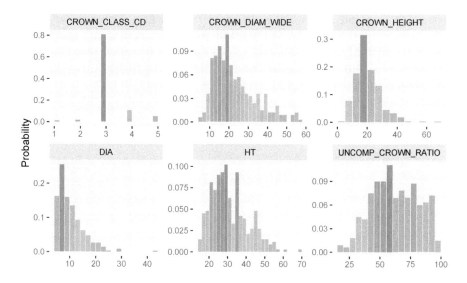

Histograms of numeric columns in df::elm2

FIGURE 15.4 Example figure from the inspect_num function in the inspectdf package that provides histograms of all numeric variables in the elm data set.

subsequent analysis. Packages like **inspectdf** allow an analyst to inspect, compare, and visualize different data sets in R. These methods can specify a data-checking routine that can help you save time, money, and resources. Identifying potential data inaccuracies or inconsistencies soon after data collection can help eliminate any statistical errors or inappropriate decisions that are made from those data further down the line.

During your statistical analysis, create and maintain a core suite of functions that you regularly apply to your data. While it is tempting to copy and paste code you used from a previous analysis (or within the same file), this can result in errors when copying, pasting, and replacing elements of code. Follow Wickham and Grolemund's (2017) advice to write a function whenever you have copied and pasted a block of code more than twice.

The tidyverse has a number of supporting packages that can help in automating your workflow. A few specific packages include:

- **tidymodels** is a collection of packages for modeling and machine learning,
- **stringr** provides a set of functions designed to facilitate the use of strings of text,
- **lubridate** provides a set of functions to make working with time and date variables easier.

For more detail, consider the **tidymodels** and **broom** packages in the tidy-verse. As we've seen, a data analyst's best friend in the **tidyverse** is the `group_by()` statement. With the **elm** data, we can fit a simple linear regression predicting HT using DIA. Adding the `group_by()` statement specifies the model for each of the five different crown classes. The `tidy()` function available in the **broom** package provides a set of functions that put model output into data frames. Here, we can see that the models fit to each of the crown classes result in a different set of coefficients and other attributes like p-values:

```
elm_coef<- elm %>%
  group_by(CROWN_CLASS_CD) %>%
  do(broom::tidy(lm(HT ~ DIA, .)))

elm_coef
```

```
## # A tibble: 10 x 6
## # Groups:   CROWN_CLASS_CD [5]
##    CROWN_CLASS_CD term        estimate std.error statistic  p.value
##             <dbl> <chr>          <dbl>     <dbl>     <dbl>    <dbl>
## 1              1 (Intercept)    10.6     0.875      12.2   6.70e- 3
## 2              1 DIA             1.08    0.109       9.92  1.00e- 2
## 3              2 (Intercept)    36.0     6.45        5.58  5.05e- 3
## 4              2 DIA             0.340   0.341       0.997 3.75e- 1
## 5              3 (Intercept)    15.9     0.973      16.3   4.29e-42
## 6              3 DIA             1.48    0.0810     18.2   8.89e-49
## 7              4 (Intercept)    15.1     3.34        4.51  7.46e- 5
## 8              4 DIA             1.69    0.407       4.16  2.05e- 4
## 9              5 (Intercept)    18.1     4.06        4.45  4.00e- 4
## 10             5 DIA             1.81    0.552       3.28  4.70e- 3
```

The addition of the `group_by()` statement allows for the same model form to be fit across different groupings of the data. From here, the **elm_coef** data set can be used to visualize the coefficients, assess any patterns in the residuals, and make predictions on new observations.

After your statistical analysis, use many of the principles outlined in this chapter to create tables and figures that your audience will value. Become familiar with the kinds of graphics that others in your discipline use to present data, and practice how to create them in R. Wickham's (2016) book on **ggplot2** has a plethora of information on designing customized themes and styles to showcase the best (and sometimes, worst) of your data and analysis. When designing tables, use packages like **kable** and others to generate and customize tables that fit your audience's needs.

15.4.3 Exercises

15.7 With some R script you have used to analyze a data set, create a new R Markdown file that produces a report that includes text, R code, tables, and figures. Knit the R Markdown document to an HTML format to view how it would appear on the web.

15.8 Use several functions found in the **inspectdf** package to examine the **airquality** data set:

 a. Use the `inspect_types()` function to determine the types of variables in the data (e.g., numerical or factor variables).
 b. Use the `inspect_na()` function to determine the number of NA values for each variable.
 c. Use the `inspect_num()` function to create a histogram of all quantitative variables in the data set.

15.5 Summary

Just like every analysis should start with visualization, it should also end with it. The choice of whether to present results from your analysis as a table, figure, or some other interactive display should be considered throughout all aspects of the data analysis life cycle. The content, construction, context, and consumption of a visualization should be considered as you are creating and presenting it to others, whether in the form of a written document or oral presentation. You can implement several practices that automate your work in R that make it reproducible and easy to implement and interpret by yourself and others in the future.

Having high-quality visualizations of your statistical results can lend credibility to your work. Devote time to learning good data visualization skills and practice them often to add more tools to your statistical analysis toolbox.

15.6 References

Essawy, B.T., J.L. Goodall, D. Voce, M. M. Morsy, J.M. Sadler, Y.D. Choi, D.G. Tarboton, and T. Malik. 2020. A taxonomy for reproducible and replicable research in environmental modelling. *Environmental Modelling & Software* 134: 104753.

Gelman, A. 2011. Why tables are really much better than graphs: rejoinder. *Journal of Computational and Graphical Statistics* 20: 36–40.

Kurnoff, J., L. Lazarus. 2021. *Everyday business storytelling: create, simplify, and adapt a visual narrative for any audience.* Wiley: Hoboken, NJ. 279 p.

Powers, S.M., and S.E. Hampton. 2019. Open science, reproducibility, and transparency in ecology. *Ecological Applications* 29: e01822.

Sandve, G.K., A. Nekrutenko, J. Taylor, and E. Hovig. 2013. Ten simple rules for reproducible computational research. *PLOS Computational Biology* 9: e1003285.

Texas A&M Forest Service. 2022. An introduction to My City's Trees. Available at: `https://mct.tfs.tamu.edu/app`

USDA Forest Service. 2022. The southern forest products industry, an ArcGIS StoryMap. Available at: `https://usfs.maps.arcgis.com/apps/MapSeries/index.html?appid=7f8429df087e4c86951a7e69d93207a7`

Wickham, H. 2011. Graphical critique and theory. Presentation slides. Available at: `https://vita.had.co.nz/papers/critique-theory.pdf`

Wickham, H. 2016. *ggplot2: elegant graphics for data analysis* (2nd ed.). New York: Springer. 276 p.

Wickham, H., and G. Grolemund. 2017. *R for data science: import, tidy, transform, visualize, and model data.* O'Reilly Media. 520 p. Available at: `https://r4ds.had.co.nz/`

Wickham, H. 2022. Mastering Shiny. Available at: `https://mastering-shiny.org/index.html`

Xie, Y., J.J. Allaire, G. Grolemund. 2019. *R Markdown: the definitive guide.* Chapman and Hall/CRC. 338 p.

Appendix

Appendix A. Statistics in R Cheat Sheet

The following provides guidance on the types of statistical tests to perform depending on the nature of the variables of interest. Several R functions are also provided, some of which are available in packages.

For quantitative variables

- If you want to compare if mean values differ from known values, perform a **one-sample t-test**.
    - e.g., t.test(x, mu = 999))
- If you want to compare if mean values differ from other mean values, perform a **two-sample t-test**.
    - e.g., t.test(x, y)
- If you want to find the difference between paired data that are not independent, perform a **paired t-test**.
    - e.g., t.test(x, y, paired = TRUE)
- If you want to find the sample size necessary given your data, perform a **power test**.
    - e.g., power.t.test()
- If you want to compare if the variances from two populations differ, perform an **F-test for variance**.
    - e.g., var.test(x, y)
- If you want to see how correlated two quantitative variables are, calculate the **Pearson correlation coefficient**.
    - e.g., cor(x, y) or cor.test(x, y)
- If you want to predict a quantitative variable using information from a different quantitative variable (the independent variable), perform a **simple linear regression**.
    - e.g., lm()
- If you want to predict a quantitative variable using information from multiple quantitative variables (the independent variables), perform a **multiple linear regression**.

- e.g., lm()
- From the **leaps** package: regsubsets()
- From the **GGally** package: ggpairs()
- From the **car** package: vif()
- If you want to predict a quantitative variable at one or more treatment levels, perform an **analysis of variance**.
    - e.g., lm(); pairwise.t.test()
    - From the **agricolae** package: lsd.test()
- If you want to predict a quantitative variable at one or more treatment levels and a quantitative covariate, perform an **analysis of covariance**.
    - e.g., lm(); pairwise.t.test()
    - From the **agricolae** package: lsd.test()
- If you want to predict a quantitative variable using information from a different quantitative variable (the independent variable) with fixed and random effect, perform **linear mixed models regression**.
    - From the **lme4** package: lmer()

For proportions

- If you want to compare if its mean proportion differs from a known proportion, **perform a** one-sample test for proportion.
    - e.g., binom.test()or prop.test()
- If you want to compare its mean proportion to another mean proportion, perform a **two-sample test for proportion**.
    - e.g., prop.test()

For categorical variables

- If you want to test if there is a relationship across categories, or see if the categories are independent, perform a **chi-square test**.
    - e.g., chisq.test()

For binary variables

- If you want to predict a binary variable (e.g., yes/no) using information from one or more quantitative/categorical variables (the independent variables), perform a **logistic regression**.
    - e.g., glm(family = "binomial")

For multinomial variables

- If you want to predict an unordered multinomial variable (e.g., three or more responses) using information from one or more quantitative/categorical variables (the independent variables), perform **multinomial logistic regression**.
    - From the **nnet** package: `multinom()`

For ordinal variables

- If you want to predict an ordered multinomial variable (e.g., three or more responses) using information from one or more quantitative/categorical variables (the independent variables), perform an **ordinal regression**.
    - From the **MASS** package: `polr()`

For integers

- If you have non-negative integers (e.g., 0, 1, 2, 3, …) and you want to predict an integer using one or more quantitative/categorical variables (the independent variables), perform **count regression** (e.g., Poisson, negative binomial).
    - e.g., `glm(family = "Poisson")` or `glm.nb()`
- If you have non-negative integers with many zeros, and you want to predict an integer using one or more quantitative/categorical variables (the independent variables), perform **zero-inflated count regression** (e.g., zero-inflated Poisson or zero-inflated negative binomial).
    - e.g., `zeroinfl()` from the **pscl** package

Index

9781032258782